William Joseph Flagg

Three Seasons in European Vineyards

William Joseph Flagg

Three Seasons in European Vineyards

ISBN/EAN: 9783337330217

Printed in Europe, USA, Canada, Australia, Japan

Cover: Foto ©berggeist007 / pixelio.de

More available books at **www.hansebooks.com**

THREE SEASONS

IN

EUROPEAN VINEYARDS:

TREATING OF

VINE-CULTURE; VINE DISEASE AND ITS CURE;
WINE-MAKING AND WINES, RED AND WHITE;
WINE-DRINKING, AS AFFECTING
HEALTH AND MORALS.

BY WILLIAM J. FLAGG.

NEW YORK:

HARPER & BROTHERS, PUBLISHERS,

FRANKLIN SQUARE.

1869.

PREFACE.

I THINK my work will be found in some degree interesting to the general reader, if he have curiosity, which Hume defines as "the love of learning."

I think, too, it may prove instructive to the general drinker as well, inasmuch as it relates to his daily beverages, and their effects on his health and happiness.

But my chief aim has been to convey information, both practical and theoretical, bearing on the important matter of wine-growing in America. Inasmuch as such information has of necessity got interwoven and somewhat entangled throughout the whole texture of the narrative, and might consequently be difficult to refer to, I have added an index, which will help the reader to search out what he may need to find, under the several heads of "planting," "training," "pruning," etc.

To the same end, I would here indicate, in advance, a few of the more important matters which will be found mentioned here and there, and not always just where they ought.

These are:

1. Long pruning, which, as commonly practiced in America, I deem to have been an efficient cause for the decay of our vines.

2. Drainage, the want of which, especially in the Ohio Valley, I feel quite certain has been equally injurious.

3. The advantage of growing wine on plains rather than on hills, except where the quality obtained from hill-grown vines is such as will compensate for their larger cost and smaller yield.

4. Training in low souche, and without stakes, as probably better adapted to our warm summers than the expensive methods imitated from countries where peaches can only be ripened on trees flattened and fastened to the south sides of high walls.

5. Red wine, as preferable to white, for the future beverage of Americans.

6. The sulphur-cure, as entirely efficacious against the disease of the vine in all its many forms, if only well applied.

CONTENTS.

CHAPTER XIX.

CHAPTER XX.

THREE SEASONS

IN

EUROPEAN VINEYARDS.

~~~~~~~~~~~~~~~~~~~~~~~~~

CHAPTER I.

BORDEAUX.

I WRITE this book because I have something to
say, and not because I have to say something.
It is of small importance how I tell what I know,
but I know what I have to tell is important. Probably no other American has made near so thorough
a pilgrimage among the vineyards of Europe as I
have, and certainly not among those of France.
When I began my explorations I had barely enough
knowledge of wine-growing to know what it was I
needed to learn, which was better, perhaps, than to
know so much as to feel above the need of learning

any more. It was only gradually and slowly, as I continued my investigations, that I became aware how much was to be gleaned from the experience of other and older countries to enlighten the inexperience of our own, and of the importance of the observations I was making, or, rather, the things I observed.

On the 20th of September, 1866, I arrived in beautiful and rich Bordeaux, the capital of wine — the centre, not of one great wine district merely, but of many, and all of them of ancient and universal celebrity. Adjoining it on the northwest is Médoc, where stand famous Châteaux Margaux, La Tour, and Lafitte. Farther away, to the north, is the department of Charente, where in every hamlet true cognac is distilled; southwardly, and up the River Garonne, lies that strip of sandy shore, narrow in measure but wide in reputation, where Yquem reigns supreme among gardens, where grow first, second, and third class Sauterne, the white rose of the Bordelais, as Médoc is the red; while all around, intermediately and beyond, are the comparatively inferior soils which yield the staple commodity of the Bordeaux market, the claret of commerce.

From the city out to the sea flows the wide and deep Gironde, the ebb tide of whose waters is a flood tide of wines, going out in ships of every nation to every port of the globe.

In the skill of the Bordeaux merchants for combining and improving crude wines I can readily believe, for what is it but chemistry and cookery—science and taste—and who are such chemists, or who such cooks, as Frenchmen? This skill has made the exports of their cellars the most portable, merchantable, and generally consumed of all the wines of commerce. For this reason they are able to market not only the product of the surrounding country, but also large supplies drawn from the south of France, taking, on an average, half the large crop of L'Ilerault, which they buy at from ten to twenty-five cents a gallon, and sell again at so great a profit, at least when Americans are the buyers, that lately large quantities were seized in the New York Custom-house upon the very natural presumption that what was retailed for six dollars a dozen on one side of the Atlantic must have cost over six francs on the other.

I ought to have made a few visits to the commercial cellars and store-houses where this great commerce is carried on, but did not do so. I was tempted away into the open country by the beautiful and soft weather which had just succeeded to the almost

incessant rains of that summer of 1860, so disastrous
to the cultivator.   Contenting myself, therefore, with
a pilgrimage to the statue of Montesquieu in the
public garden, and a short stroll on the docks, I got
into an omnibus running to Créon, in the neighborhood of which was the chateau of a gentleman
whom, though I had only met him once, and six
years before, I was resolved so far to impose upon as
to ask leave to look at his vines.

"Why there's one of Rosa Bonheur's horses!" I
exclaimed, as a dray went by the omnibus, drawn by
one of the larger specimens of that admirable race
of animals which the well-known engraving of the
horse-market has made familiar to us all, on paper.
"Is it possible those casks are full?"   "Yes, sir,"
replied my neighbor on the opposite seat; "and there
are fifteen of them, each holding 228 litres, and with
the wood weighing good 250 kilos.   But they can
well do that, those Normandy fellows — beasts of
nerve they are."

And inquiries repeatedly made while I remained
in France satisfied me that it was indeed possible for
the heavy draft horses of Normandy to draw on one
of the enormous drays that are made for them between three and four tons.   If we could replace our
six millions of nags, of one sort and another, with

one third their number of a breed like this, the two millions would do the work of the six, at a saving in feeding and attendance equal to double the interest of our national blessing. Thus computing, I said to myself that if I had left behind me the land of steam, I had found the land of horses.

Two farmers, whom I afterward met while traveling in Normandy, told me the Perche country was really the home of the breed called Norman, and that it was their custom to buy from there six-months' colts, which they raised and broke, working them from two years' old, and selling them when they got to be five or six years old; the prices obtained for full-grown and well-broken animals ranging from $200 to $250. I am glad to learn they are at length bringing them to America, where a late importation sold for prices which averaged $2500.

Falling into conversation with my fellow-travelers, I was gratified to learn that M. P——, whom I was going to see, was esteemed a skillful and successful cultivator, with few equals in the neighborhood. I talked a good deal with my companions in the omnibus during our two hours' drive. They were mostly working vine-dressers, and being, as Frenchmen always are, polite and communicative, I learned from them a good deal I had never heard before concern-

ing the object of my inquiries, if one can be said to
have learned any thing when the lesson he takes is a
confused jumble of details which overloads his memory and befogs his intellect.

From my peasant acquaintances collectively I did,
however, obtain the following clear ideas:

1st. That each variety of vine needed a different
culture for each different soil, and again for each
different climate.

2d. That there was an old school and a new school,
with opposite theories on every branch of vine culture, planting, manuring, training, pruning, cultivating, and gathering.

3d. That every cultivator had his own whims and
prejudices to qualify his application of the newer theories.

4th. That there were a good many varieties of
French vines, and a good many different soils and
situations in France.

From all which I inferred:

First, that there was much to learn.

Secondly, that I should never learn it.

But a few clear lights will illuminate a good many
facts, so that with patience and labor the rubbish can
be known and rejected, and the useful brought into
form and order.

CHAPTER II.

SAINT GENES.

THAT night I lodged in the only inn at Créon, a humble little affair where the peasantry resorted to enjoy their hard-won leisure and drink their wine, but where the food and bedding were good enough for any body. The next morning I was driven over to the château of St. Genes, whose proprietor recognized and welcomed me with the politeness of a Frenchman and the hospitality of an American.

With small loss of time, and without needing to go far, we began the tour of M. P——'s well-kept and extensive fields. Having long attended to his own affairs, he was well informed on every practical detail; and having once been a lawyer, he could explain them fluently. The weather was fine, the country was beautiful, and I was happy to be walking in a French vineyard that day.

The soil of the first piece we entered was a sandy loam; in other places I found it to be gravelly loam, but all was mixed with more or less of clay. The better wines grew on the gravel. The piece in question was furnished with wire trellis. The vines were set two feet apart in the rows, and the space between the rows was four feet wide. The posts were round and straight locust saplings grown for the purpose, and were placed twenty feet apart. Through holes in them the wires were strung, and an ingenious contrivance tightened them. They were further supported by intermediate stakes. There were three lines, each being eighteen inches from the other, and the lowest at the same distance from the ground.

The fruit-bearing cane was trained along the lower wire, so that the bunches seemed to belong as much to the one as the other. The canes thus trained, however, are not allowed to grow into arms, but are renewed every one, two, or three years. The fruit of the second year, and which was produced from buds on the shoots grown during the first year, seemed to hang so close to the horizontal cane and wire that I think those shoots must have been cut back to one eye only, but on this point my recollection is not quite distinct. The shoots, as they grew, were attached to the two upper wires. Although the season had been bad, the grapes were healthy, and with a fortnight or so of the fine weather just set in, promised to do tolerably well. M. P——'s wire trellis was indeed a pretty sight. That gentleman thinks the wire saves one half the cost of manipulating-the vines; namely, of training, pruning, attaching, rubbing off, pinching back, unleafing, amd gathering.

"What is that?" I exclaimed, with no little astonishment, as, turning away from the trellis where vines were so tenderly upheld, we entered on a field where there was never a bit of trellis nor stake at all, nor peg to tie to, nor tree to hang upon, but where each individual plant, alone and self-sustaining, scorning all support—its arms embracing nothing, its tendrils twining nothing—stood on its own bottom, and held up its own top, like a strong-minded woman planted on her rights!

It was a field of the variety known as "*la folle blanche*" (the crazy vine), vulgarly called "*enragatt*," growing "*en souche basse*," which may be translated by stump or stool, *souche* meaning literally "*stock*."

I paused long in presence of this abrupt commentary on all our learned talk about different kinds of trellis and modes of training to them, and did not move on till I had learned something about training "*en souche*" and "*la folle blanche*."

I learned it was an uncommonly hardy plant, nev-

or injured by frost, nor, to M. P——'s knowledge, by any disease; that it was a regular and reliable bearer, and, on a good sandy loam, such as I then saw, could be counted on for over a thousand gallons to the acre, and sometimes gave as much as twenty-five hundred.

As a workman drew apart the branches of one of the souches, a profusion of full-sized white grapes was revealed, all hanging close about the head, and easily sustained by the rugged old stock, which was about ten inches high and five inches thick. "It is a perfect fountain of wine," said the man.

The quality of the wine from the folle blanche depends, of course, much upon the soil. In Médoc they habitually grow it with the malbec, a fine variety, but whose must is deficient in acid; and the combination results in a wine of the very first grade, such as sometimes sells from the cellar at eight dollars a gallon. Although commonly grown for quantity, and on strong soils, it nevertheless makes the most of its advantages, and on gravelly loam will give a very good merchantable white wine. The Bordeaux merchants compound it with a strongly-colored coarse wine from the back country, costing twenty-five and thirty cents a gallon, to make a cheap claret, which is sold, labeled with the names of all the great houses of Médoc, to Americans. The price of wine from the folle blanche is forty cents and upward, though M. P—— sells his for fifty and sixty cents and upward. In the department of the Charente this plant is the favorite, and chiefly from its strong juice the Cognac brandy is made.

Now white wine mixed with red does not make a true red wine, and those of us who drink such compounds as the above drink two distinct beverages mixed together. But both are pure, and, if not adulterated with alcohol, wholesome. Delavan and Dow tell us that all our imported wines are not wines at all, but mere chemical illusions, as if France, with a yearly product of a thousand million imperial gallons, needed to draw upon her cisterns, wells, and drug-shops for the small quantity she exports. In some parts of the south they sell pure wine at wholesale for a cent the bottle—not very drinkable stuff, to be sure, but a good deal better than dye-stuff, one would think, and cheaper too.

M. P——, seeing how much his wine-fountains interested me, kindly offered to send me some cuttings from them. Knowing how completely had failed all attempts to acclimate European varieties in America, I did not then accept the offer; but a few months later, and after witnessing in the south of France the wonders of souche training, I reconsidered and accepted. He sent them. They arrived safe, a thousand of them, whereof three hundred took good root, and are now growing finely on the banks of the Ohio.

Now when I shall come to relate my observations in the south of France, my reflections thereon, and plans and hopes thence resulting, I think they will be found new, interesting, and important to my fellow vine-dressers. I think they will see in souche training the true way to get wine cheaply and easily, so that none shall need to drink water, except, as Fortescue, the chancellor of Henry the Sixth, wrote of the common people of England in the days when she was "merrie," "occasionally, or by way of penance." And in the day when every farmer can, from half an acre of land, easily and cheaply planted and tilled, even by the unskillful, harvest what will fill his ten or twelve barrels with honest juice for the habitual daily drink of himself and family—our two heavy afflictions and sins, excessive water-drinking and excessive whisky-drinking, will vanish from the land, and a beneficent change in our national temperament begin to be wrought.

The vines about Créon are not generally of so low a class as the folle blanche, neither do they give great wines, such as are made in Médoc or the Sauterne district, but are of those rather which yield the good, staple "Bordeaux," dearly loved of all Frenchmen, and for which they must pay no very moderate price either, since much of it commands, at wholesale, a dollar a gallon. It can be had in America of honorable wine-merchants dealing with others like themselves on the opposite side of the water, or, better still, who have direct relations with honorable proprietors there who reside on their estates.

The fields we next inspected were in good cultivation, but the vines were trained to stakes only, reminding me of those in the vineyards I had left behind, except that they stood nearer together and were rather smaller. They seemed to have had no very close summer pruning, but little tying up, and no leaf-pruning, though the time for it had passed. The ground had been only twice plowed, I think.

The labor is to a great extent done by contract, and of necessity it is carefully classified and specified. Where the superintendence is good, the system works admirably. It is desirable we should introduce it as soon as the vine-culture shall be well enough extended, organized, and understood, but for the present I should fear to try it. I remember that in Brown County, Ohio, they once had, and may have still, a simple plan of letting the whole labor, by contract, at forty or fifty dollars per acre for the year, which was at the same time costing me as much as seventy-five dollars. Wages in the neighborhood of St. Genes were forty cents a day in summer and thirty in winter. Women got but half as much.

As paper money has of late years confused our ideas of values, I will in this connection give some of the retail market prices customary about Bordeaux, so that the value of thirty and forty cents may be somewhat estimated.

> Beef and mutton, good cuts..................20 cents.
> Pork...14 "
> Eggs, per dozen................................12 "
> Milk, per quart................................. 3 "
> Strong shoes, good for a year's wear, $1 60
> Wooden shoes.................................26 "
> The same with leather uppers..............60 "

Our promenade extended far beyond the domains of St. Genes, and over those of several neighboring proprietors. Entirely new to me and to my feet was this going from field to field, and farm to farm, as one may do in nearly every civilized country except Britain and America, without ever meeting fences to be climbed, walls to be scaled, bars to let down, or gates to open. The contrast between the orderly, neighborly, and trustful aspect of the scene I was studying, and the fortified look of our own cultivated country, where at every few rods you encounter picket, or palisade, or barricade of stone, or double stake and ridered nine-rail worm fences, bristling like so many abattis, all of them "pig tight, bull strong, and stallion high," was like the contrast between peace and war.

Returning rather late to the chateau, we could give only a few moments to the wine-house. I was pleased to notice a hand-mill for crushing the grapes—a good deal nicer way than what I saw a few days later among people less advanced than my host of St. Genes. He told me, upon my inquiry, that the crop of the estate the year before—an extraordinary good one—was 500 barriques, or 30,000 gallons.

At dinner I met the ladies of the family, which, had I done before my walk, it would have been shorter, perhaps. M. P—— resides in Bordeaux, and the family had only come out to Saint Genes to remain through vintage. He, however, having a business-like way of looking after his interests, is frequently there.

Next day my good friends would not allow me to go back the way I came, but drove me over to a railway station some ten miles distant, the drive affording a sight of extensive vine-fields, and some most charming scenery as well.

CHAPTER III.

COGNAC.

THE speed of common railway trains in France
never takes away your breath, nor whirls things
out of sight before you see them. So nothing hindered my observing all we passed, on both sides of
the track, leisurely enough to get an idea of the
modes of training, and so forth, in the northern portions of Aquitaine—the ancient and original—the
Aquitaine of Froissart's Chronicles. Many a vineyard I saw whose fresh young shoots and foliage
covered and hid short, thick, rugged old stocks below, gnarled and wrinkled with a hundred years of
fruit-bearing existence. But a century is a short
time in the history of the vine in this antique country. The Cæsars drank the juice of its soil and were
glad. The savage Visigoths, in their turn, we may
be sure, got beastly drunk on it. The pious Saracens
who drove out the Visigoths broke the law of the
Prophet in its honor. And from the time of William

the Conqueror down to the Methuen treaty, which excluded it in favor of Spanish adulterations, it nourished and strengthened the best blood of England, a good deal of which same blood was again and again poured out on this same soil, in battles fought to hold and extend possessions which yielded to the thirsty islanders what Nature had denied them in their proper home—good wine and red.

A fanciful theorizer has said that all good English comedies were written before the time of the Methuen treaty, which was about a hundred and fifty years ago, arguing thence that only pure wine can inspire pure wit. It is very true that both England and America mostly import their good plays from France, in shape of translations or adaptations; but I can hardly believe it was ever possible to import· them in the form of casks of Bordeaux or bottles of Burgundy; and think, with Sir Emerson Tennent, that British palates have always craved what was mixed, muddled, and strong, which, he says, is because of fogs. If so, then so much the worse for the British, I say, and all the more shame for us, who, with no fogs to excuse it, have, from mere force of example, learned to love fog medicine—port, sherry, Madeira, whisky, and rum—which, in our dry climate, rend us as they never do a Briton in his home—but

B

this reminds me we are on the way to the land of brandy.

A clattering of plates and glasses called my attention to the party occupying the same compartment with me, consisting of a gentleman and wife, two other ladies, and two children, who were beginning their midday breakfast. The bottles, of the size the Bordeaux people use when they drink, but not when they sell, held as much as one of our quarts. That family emptied one bottle, they emptied two, they emptied three. To tell the truth, they ate as beautifully as they drank, managing to divide the contents of their enormous lunch-basket into nine or ten courses, and, by taking them in detail, to conquer them all. At the end of the repast a bottle of brandy was produced, and a paper of sugar in lumps. A very little of the brandy was poured into a glass, and each one taking a lump of sugar, soaked it in the brandy and ate it. Still another bottle was uncorked, and from it a tumbler was filled with water —the first appearance of that liquid on the scene. In the tumbler they all washed their fingers and lips.

That is the way French people drink wine, and that the way they drink brandy, and that the use they make of water.

Leaving the cars at Angoulême, I continued my journey in a diligence. The fancy pleased me of traveling in the old slow coach of slow old times, as Sterne did when he made his "sentimental journey through France and Italy." But sentiment was not curled hair, and could neither cushion the hard seats nor deaden the rattling din of the rackety concern, and I was glad when they set me down at a snug hotel in the little city of Cognac.

As we entered the brandy district, the folle blanche appeared and soon covered the whole face of the country. The soil was mostly stony, poor and thin; of no value at all, I should think, except for grapes, and even a grapevine, one would think, must be crazy to live there. I could nowhere see that stakes were needed for supports, though the souches were eighteen inches high. Young plantations, I have been told, need small stakes during the first two or three years, but I noticed none.

It rained continuously. No vintaging could be seen, and attention soon tired of the look-out at the window of the coupé, so I looked in. Two nuns were with me — not handsome, of course, for in France beauty is too precious a commodity to shut up in convents—but very jolly and talkative, and better company than the holy women I had met in

America; and, though their ideas were limited
enough, still a seeker after wisdom could learn a
good deal from them. When I told them the great
majority of my countrywomen esteemed it a sin to
take a drop of wine, they were astonished, and one
naïvely asked, "Must they drink only beer, then?"
adding, "I don't like beer." But when told beer too
was forbidden, they fell to pitying the poor Protest-
ants, whom they had not thought were so austere.
"To be sure," they said, "one must do penance; it
is for the safety of the soul; but the good God does
not require his creatures to injure their health by
their abstinences."

Abstinences! Poor girls! If a marquis with
200,000 francs of income, young, handsome, and
agreeable, were to offer himself to either of them,
she would abstain from him teetotally, with might
and main, as if on peril of perdition, yet she could
put her quart of wine daily under her corsets, and
thank God for it in her prayers; while many a pretty Puritan on our side, taught from childhood to believe it "liquid poison" to body and "liquid damna-
tion to soul," thinks it a sin and a crime to moisten
her red lips with one drop of purest Margaux, on
whose conscience a hundred warm kisses accepted
by those same lips would rest as lightly as a thistle-
down on Plymouth Rock.

Arrived at the hotel, I found a seat by the kitchen fire more agreeable than imprisonment in a bed-room. The kitchen was large, and was, in fact, the chief rendezvous for all the household, as well as their guests. In my time I have stopped at many an American country tavern, and sat in their bar-rooms while my fellow-citizens came and went and drank whisky. The scene I witnessed at Cognac was, quite different. About a table in the middle of the room were seated eight or ten peasants and town-folks, re-freshing themselves with bread and cheese, and strong draughts of weak wine, while amusing them-selves with cards, conversation, pipes, and snuff. In the adjoining room was a billiard-table, where a larger party were engaged in playing or looking on. These, too, had their potations. During the two hours I re-mained below I noticed closely the conduct of all the company, and, though there was plenty of gayety and seemingly real enjoyment, there was nothing in the least like drunkenness, ill temper, or ill manners. Lager beer, so called, is an immense improvement on rum and whisky—thanks to the good Germans who have made us to know it—but simple wine certainly has moral qualities far superior to beer. A merry but decent drink, exhilarating but not infuriating, it carries neither knife, revolver, nor slung-shot in its

pockets. To the poor work-people of France it is
an inestimable blessing, as it will be to ours when it
is vouchsafed to them.

"You don't bake your poultry, then?" I said to the
landlady, as I saw her fix on a spit the fowl I had
called for, and then set it to turning before the fire
by means of a clumsy clock-work. "No, no; that's
the way they will spoil your pullet in the great hotels at Bordeaux, and then make you pay three prices
for it. For my part, I say, '*Vive la broche!*'" (long
live the spit!) The spit did look long-lived, rather;
it measured good seven feet. The fowl was roasted
well enough, and ate well enough; but it was without dressing, was soppy from frequent basting with
only water, to keep it from burning in the blaze of
fagot-wood, and came before me with its head and
claws on. Nobody in France roasts any better than
this.

In the morning it still rained hard, and I did not
care to make any excursion into the surrounding
country, where there would be no distilling to see,
because it was vintage-time, and no vintage, because
it was raining. A few years before I had received
a visit from young Mr. Otard, of Cognac, of the firm
of Otard, Dupuy, & Co., so thought I would look
him up; but, on calling at the place of business of

the firm, which I could easily see was an immense concern, I was told the gentleman in question was absent from town. Unfortunately, no other member of the house was in Cognac, and the highest authority to be found on the premises had no authority to admit a stranger.

All the world have heard of the house I have mentioned. Its name is often used in America to christen whisky. O., D., & Co. are not distillers, however, but, like the other large houses of Cognac and Jarnac, are merely merchants who buy up the liquor distilled by the country proprietors, and gather it into their magazines, where they treat it—or maltreat it—in some dark mysterious fashion they fear to let strangers witness, and, when it is old enough, sell it under their own brands.

Cognac brandy is not cheap, even in its own city, where such as is old enough for drinking costs, from first hands, two dollars a gallon. Brandy is made in many other parts of France for about half that price. It is a pity we in America must pay so excessively as we do for French brandy, and even then be tormented with doubts of the genuineness of the medicine we take.

Good physicians say the aromatic quality of brandy gives it medicinal virtues different from those of other kinds of spirits, and, moreover, that pure liquors are better medicines than adulterations. Whether any one can explain why this is so or not, I am sure no intelligent person would have equal faith in a mixture of common druggists' alcohol and water, as a remedy for typhoid fever, as in pure Cognac. There must, in the nature of things, be a connection between aroma, savor, taste, and digestibility—between attractiveness and usefulness. They brought me but lately a saddle of venison from a deer that had been chased into the river and there killèd. It had utterly lost all taste, and could not be eaten, or, had it been eaten, it would have failed to afford the least nourishment, if Liebig is right. The nervous power had been hunted out of the poor beast, and with it had been expelled from the flesh all that could be attractive or useful to man. Against whisky, as whisky, I have no objection; but as brandy, whisky is a failure. To convert it into Cognac, they first rob it of its corn ethers, and then replace them with concoctions which may cheat the palate, perhaps, but never the stomach. The connection between the ethereal and the substantial parts of all drinks is like that between spirit and matter—once dissolved, it can never be restored.

To make a gallon of Cognac brandy, seven and a

half gallons of wine must be distilled. No sooner
has fermentation subsided than distillation begins,
and this is often as early as the first of September.
Three qualities are made in the Charente: great
champagne, little champagne, and bois. The term
champagne comes from the resemblance of the soil ·
where the wine is grown to that of the department
of the Marne, in the province of Champagne, both
being chalky limestone. The best quality is from
the poorest soil, of course. The average yield of
wine to the acre is 400 gallons. The cost of cultivation is about twenty dollars, gathering and pressing
included. Plowing is done four times a year, twice
to uncover, and twice to cover the feet of the souches.
A regular and certain return of five per cent. on his
capital contents the proprietor in the Charente, and
even this moderate rate could not be realized but for
the use made of the space of twenty feet left between the rows for raising general crops.

B 2

CHAPTER IV.

MÉDOC.

THE district named Médoc lies to the northward of Bordeaux, with the River Gironde for its eastern and the ocean for its western boundary, and is a peninsula of considerable extent. But the valuable portion of it, called "Haut Médoc," is but a narrow strip, not more than a mile and a half wide, nor more than thirty miles long, occupying the slightly raised middle ground between the sandy and sterile "landes" of the sea-coast and the rich alluvial ground of the river border. Beyond question, this narrow belt is the most notable piece of all the earth's surface for growing red wine. The reader and I are going there to-day, not for any purpose of amusement, but on the important business of learning how to make red wine, we and our countrymen being as yet alike lamentably ignorant of it.

Yet it is what we must needs know something about, for the wine of our future must be red, and

not white. To Médoc we will go to receive our first
lesson, nor could a better school be found beneath
sun or moon.

For my part, I will emulate Pliny when he said,
"I shall discourse of wine with gravity becoming a
Roman treating of useful arts and sciences, approaching my subject, not as a physician, but as a judge,
who is to pronounce on the physical and moral health
of the human race."

A little deep-draft, narrow steamer of sea-going
model, whose small spluttering wheels turned swiftly
enough, but to wonderfully little purpose, conveyed
me down to Pauillac, and was all day about it. But
what if it did go slow? it carried me safe and returned me sound. I don't know why we should suffer ourselves to feel contempt for the small craft of
European rivers and lakes. Narrow and sea-sicken-
ing as they are below deck, cramped and shelterless
as they are above, they are arks of safety to life and
limb, and an improvement on Noah's, I dare say.
True, the two boats on which I used to go and come
between city and country home could either of them
singly do the whole business of the Upper Rhine, the
Gironde, or any Swiss lake, without drawing more
than three feet of water, yet both Boston and Bos-
tona, the one about the time I am writing of, and the

other three years before, took fire and burnt up while under full steam, making excellent speed, and full-freighted with passengers and goods.

But, though it took all day, the day was not lost. There was a good deal to see. We were continually passing inward-bound ships of every nationality, riding at anchor till the flood should come and tide them up to the city, there to discharge their varied cargoes and again reload, two in every three of them with claret and cognac. One of them bore the flag of my country, and as I gazed on its folds I knew it would soon be waving proudly over a homeward-bound cargo of as inferior liquor as Bordeaux could export.

The deck was crowded with people, mostly of the peasant class, and all of them going to vintage. The freight piled up forward, casks, baskets, and queerly-fashioned tubs, was going to vintage too, and every thing spoke of festive labor. The men wore blouses, mostly of blue linen, and the women had only caps for bonnets; yet really both men and women seemed to me better dressed than the working-people I had left at home. Perhaps this was less owing to the quality of the stuffs worn, though few were poor enough to appear in calico, than to the fitness of the costumes for the daily avocations of the wearers.

Then.for their deportment—I don't know how they would have appeared if translated to the saloon of fashion—awkwardly enough, perhaps; but, taken as they were, in their habitual sphere, the manners of those Bordelais peasants were such as our people can never emulate, I fear. They were, in a word, respectful and ceremonious, yet natural and easy; graceful, yet simple; gay and talkative, yet quiet and reposed.

French theorists have claimed, be it known, that although a select class of English or Russians may, by mere dint of high breeding, become civilized and refined, yet the masses of their fellow-countrymen, as well as of all peoples who are without wine, must forever remain barbarians. If there be any thing in this theory, I would prayerfully entreat the Genius of Civilization, or the Spirit of the Age, or god Bacchus, to take up bodily the whole American people, men, women, and children, youths and misses—especially the youths and misses—and plunge us all up to the lips in a sea of the proper liquid, therein to soak and thereof to swallow, until politeness shall penetrate all our joints and muscles, and refinement enter into the texture of our bones.

There were some oysters on board—Médoc oysters, of great repute through France, as were their ancestors among the Romans. As early as the fourth cen-

tury they were mentioned in somebody's writings as
"a shell-fish as much esteemed on the tables of the
emperors as were the excellent wines brought from
Bordeaux." I tasted them in as impartial a mood
as Pliny's, as I afterward did the other celebrated
kind brought from Ostend, but in neither could I
find any excuse for Roman gluttony, nor any thing
else worth swallowing. Watery, thin, and coppery
are Europe's best oysters, and watery and fishy are
her worst.

Near the close of the day we arrived at Pauillac,
and I found out and entered a little old inn in the
heart of a labyrinth of narrow streets, where a tough,
chirpy old woman received me as if she had always
known and long been expecting me. She seemed to
know just what I wanted to learn, and, having shown
me the chamber where I was to lodge, and the parlor
where I was to eat, took me to the kitchen, and displayed the preparations she was making for a band
of vintagers soon to come in from their work in the
vineyards—for she was herself a proprietor, it seemed. Lifting the lid of a large kettle, and letting me
smell of a savory mess within, she told me it was
the vintage broth, a dish of great antiquity. Judging from the preparations, the vintage band was a
large one; in fact, she had a farm of no small value

—was rich, in fact, but it was her humor to keep tavern. The proprietors, she said, always fed the bands of vintagers, and gave them three repasts daily; the first, at eight o'clock, consisted of only bread and grapes; the second, in the field, at noon, of soup and soup-meat; and the last, in the evening, of soup and a ragoût of meats. Bread and wine were supplied "*à discretion*," which means without stint. The wine is made of inferior grapes, gathered from young vines usually, crushed and put into a barrel with the head out. As soon as fermentation has well begun, a certain quantity of water is poured in, the cask is tapped at the foot, and the liquor placed at the discretion of the drinkers. One franc a day is paid to the cutters, as those are called who cut the grapes from the vines, and for such as carry them to the wagon a franc and a half. But a few years ago wages were one fourth lower.

After my dinner I had grapes for dessert, and they were choice bunches, and good as any I afterward tasted in Médoc, but I could not call them delicious. Nor did I any where in France, Switzerland, or Germany, find any to equal the Catawbas of the Ohio Valley in their prime.

In the evening I went to see a vintage dance at a chateau just outside the town. Under a shed, lighted with a single candle, twenty or thirty of the younger vintagers were dancing in wooden shoes on the bare ground. The figure was simply the old pantaloon cotillon of "forward two," "cross over," "right hand left," "dos à dos," and "ladies' chain," only the couples were placed in two opposite rows, as in a contra dance, and not as in a quadrille, so that the dancers were continually in motion. Occasionally this was varied by a few rounds in waltzing order, performed with a kind of balance step, the partners holding hands and facing each other. They did not hug, as fashionable people do, nor was there any rudeness, or romping, or boisterous conduct of the men, and far less any sign of drunkenness.

A hurdy-gurdy, played by the overseer of the troop, was all the music they had. The overseer is a kind of middle-man, who recruits the band in the neighboring and poorer districts, and conducts them from place to place while vintage lasts, sub-letting them at a price which yields him a profit of two or three cents daily on the labor of each person.

I had seen vintage dances before this, at the theatre, but there was always a row of brilliant foot-lights, and a large orchestra, and the dancers wore blue and red bodices, with clean chemises, and broad straw hats adorned with gay ribbons, and had neat slippers on their feet, and pink stockings on their legs, quite unlike any thing to be seen in the stable-yard at Pauillac. One of the wooden shoes, flung from a maiden's foot as she whirled by in a waltz, struck my knee with centrifugal force. As the Cinderella who owned it kept on with her dancing, I had time to examine the "sabot." It was nicely made, of good shape, and light, furnished with a simple leather "upper" nailed to the edge of the sole. The cost was only seventy-five cents the pair. Many people in France, who live in the country, wear this kind of shoe in muddy weather from choice, and thus avoid many a malady. An American, with common sense enough to adopt them for himself and family, could save sufficient between the birth and coming of age of his oldest child to buy a farm.

"Here is where they sleep," said my guide, as he stopped before an open door. I looked in, and saw merely a large room in an out-building, the floor of which was covered with a comfortable thickness of clean straw, upon which straw some forty vintage youths and maidens were to sleep that night. This, they told me, was the usual mode of lodging the laborers. They seemed very happy—and why should not they be? those trooping bands, tramping from one merry harvest to another, seeing the world for

nothing, free for the time from home restraints, fed
and lodged in foul weather as well as fair, earning
wherewith to buy clothes for the year by light, social, and agreeable labor in the day, and enjoying a
vintage ragoût and vintage dance in the evening, eating "à discretion," drinking "à discretion," and sleep-
ing "à discretion."

When I went down to breakfast the following
morning, I found madame was already up and a-field,
having left her only domestic to attend upon her only
lodger.   Mathilde informed me M. Averous had been
to call on me, and had left his compliments, with the
offer of his services to conduct me to see the vintage.
It was the landlady, I learned, who had obtained for
me this polite attention.   To lose no time, I waited
on myself while Mathilde ran to bring a hack.

Thanks to a soft, fair, welcoming kind of weather,
such as makes you feel at home in a strange land, I
could go in an open carriage.   French towns take
small space, and in five minutes I was beyond the
outer borders of Pauillac, and going along a vine-
bordered country road, where, for leagues on either
side, nothing hindered the view.   Soon we began to
pass wagons loaded with fruit on its way to the vats,
each drawn by two oxen of a most noble breed.
Their color was a tawny drab, and their horns white.

They seemed thoroughly trained, and moved along in a dignified manner, as if they drew their load of their own free will, and not from fear of the slight rod armed with only half an inch of darning-needle, carried, rather as a guide than a goad, by a man who walked beside them without blasphemy or loud words of any kind. It means something that the French use the word "conductor" where we say "driver."

Every ox wore a net over his face—quite a neat thing, too—and a cloth that covered the back and hung down to the knees, which were for protection against insects such as swarm from the low lands of the river border. This highly-esteemed race is the result of kind and judicious treatment, as much as of the rich pastures of the Gironde.

In Ohio I could never get an ox-driver to undertake any heavy work without a fresh sea-grass snapper at the end of his short-handled, rattlesnake-looking whip, nor unless his own lungs were in good order for swearing. In one year I received the resignations of two good drivers, tendered solely because their lungs had given out. Both were good men, and really meant nothing but business when they swore and scourged. What breed of beast such evil influences and rude discipline will produce the future will reveal.

It was Fourier who taught that, so soon as mankind shall learn to take good care and make good
use of the domestic animals they already have, the
Creator will give them others more perfect and more
useful. I don't know what authority the philosopher
had for this promise, but am sure the ass-drivers of
Naples and ox-drivers of some parts of America will
have to wait a good while yet for the prize animals
which are to reward their humanity, while one might
fancy that in the Percheron horse and oxen such as
I have described, the French people had already received their recompense.

The large group of cutters to which I was directed to find the Messrs. Averous evinced that those
gentlemen cultivated on no small scale. One of them
came to the carriage to receive me, and I soon found
myself at home in the busy company, and fell to eating grapes and asking questions, the first one being
why were the vine-leaves so spotted in many parts
of the field? It was verdigris, that had been sprinkled on the outer ranges of vines to keep away birds.
Whether it poisoned or frightened them I now forget; but a bird that comes fluttering about and drinking, "à discretion," wine worth a dollar the bottle,
without a cent in its pocket, is a sponge, and deserves
verdigris.

The organization of the vintage troop I found to be quite systematic. First there is the rank and file, mostly women and children, who go along between the rows, one in each space, and gather the fruit into tight baskets. These cut off the bunches with knives, and are called "cutters." Following within easy reach of the cutters, along alleys which cross the rows at suitable and regular distances, go the wagons, each containing two short upright casks without heads, and drawn by a yoke of oxen. Between these and the cutters come and go men who carry on their backs, with the help of shoulder-straps, the common deep, oval tub, of size to hold five baskets, such as the cutters carry, and called "hotte." The hotte-bearer has in his hand a stout walking-stick, which serves to prop his burden so as to relieve him of its weight while standing still waiting on the cutters. His vessel filled, the hotte-bearer carries it to the wagon, mounts it by a short step-ladder, dumps his load into one of the casks by a quick inclination of the body, and then, with his stick, stirs about the grapes to pack them well down. Over all is the "commandant," whose name implies his duties. His insignia of office is a long slender lath, to the end of which is fixed a willow twig. When a cutter commits a fault, such as leaving a bunch ungathered, or

not properly culling out the green or decayed berries, instead of calling it to her notice by words,
which would draw the attention of the whole group
and cause a considerable loss of time, he lightly
touches her shoulder with the tip of the twig.

Every twelve cutters had two hotte-bearers and
one commandant. Such a force will in a day harvest a "hectare," about two acres and a half, bearing the average yield of Médoc, which is 625 gallons, or 250 to the acre—a very large yield, considering the fine quality of the wine. This is good
work, but they go early to the field, and never wait
for the dew to dry from the fruit, as is common elsewhere, since they do not fear it will do any harm in
the vat.

The rows were three feet apart in all the Médoc
vineyards I saw, and the vines the same distance
from each other in the rows. They stood on little
ridges flung up against them by the last of the four
plowings which they annually receive, two of which
uncover, and the other two cover their feet.

The plow they use is of wood, except a plate of
iron in form of a long triangle with which it is shod,
and has but one handle, being, in fact, no other than
the Roman implement of Virgil's time. It is drawn
by two oxen yoked abreast, one of which treads in

the space between the rows where the plow is moving, and the other in the next one. The beam is curved quite curiously to insure the proper bearing and direction, and must of course pass above the tops of the vines, stakes, trellis, and all; and, strange to say, all of these are kept within the low stature of fifteen inches for no other purpose than to allow the plow-beam to pass over them. If there were ever any other reasons for this Liliputian training, this is the only one that has come down from the remote antiquity which clouds the origin of the customs of Médoc.

One of the gentlemen took me to see the wine-making in a large old stone building near by. On entering the spacious and high press-room, the first thing to see was a circle of workmen engaged in stemming the grapes. They were standing within a shallow box, or rather a wide platform with a low rim, built up about three feet above the floor, very much resembling the dish of a common wine-press, but quite large, measuring ten feet each way, and called " pressoir." Its bottom pitched a little on one side, and was grooved to let the must flow freely away; and in the rim of the lower edge was an opening about a foot wide, beneath which was a large tub of 200 gallons' capacity, to receive the

liquid, called a "gargouille." The men were at work
about a sort of table which stood in the centre of the
"pressoir," having for a top a grating or screen made
of rods of half-inch iron, and bordered with a low
rim of wood. The rods running the shorter way
were supported in the middle by passing through a
bar one and a half inch by half an inch, but those
running the longer way were fastened only at their
ends, and rested on the others rather loosely. Close
to the pressoir was a doorway opening into the yard,
at which a wagon was being unloaded as I entered,
which was done by bringing the casks from the wagon directly through the door and on to the floor of
the pressoir, which last was on the same level with
the sill of the door and bed of the wagon, so that no
lifting need be done. The contents of the casks
were dumped close to the table or screen. Then,
having flung a few bushels of the grapes upon the
screen, the workmen took their places about it and
began to rub them on it with their hands, the berries
passing through the meshes, and leaving the stems
behind. Soon currents of red juice crept out from
beneath the mass of crushed grapes under the table,
and, flowing along grooves in the floor of the pressoir, ran out through the opening in the rim and into
the gargouille beneath. It needed but a short time

to make an end of the wagon-load, and then the table was set on one side, and the heap accumulated
beneath and about it was shoveled out, by way of the
same opening which the juice went through, into
tubs made of barrels sawed in two, with sticks passing through holes bored in the staves, and projecting
on either side, for handles. As these were filled, they
were carried up to the top of the vat and flung in.
The juice accumulated in the gargouille was carried and poured into the vat by means of the same
tubs.

Climbing by a ladder to the level of the rims of
the long row of vats which lined one side of the
press-house, I could see that two of them were full,
and the contents already fermenting, covered with
only a thick float of stems.

The vats were of oak, iron-bound, eight feet deep,
ten feet wide at the bottom, and nine at the top. The
hoops were not riveted, but were clasped where the
ends met by short screw-bolts passing through flanges
or ears, the bolts serving to tighten the joints of the
staves when necessary. Each vat would hold four
thousand gallons.

No other crushing was given to the grapes than
what they necessarily got in being rubbed through
the meshes of the screen.

C

On taking my leave, I received valuable instructions how to shape my course in my proposed circuit
among the great houses of the canton, and also what
was very pleasant—an invitation to dinner—which I
incontinently accepted.

After quitting the Averous farm, my course lay
along a wide, gently-swelling ridge of gravelly land,
and commanding an extensive view in every direction. The face of the country was rolling, and divided into long, low swells of high ground occupied
by vines, and wide intermediate flats hardly above
the level of the Gironde, partly devoted to grain and
partly abandoned to a coarse pasturage. These flats
render Médoc unhealthy, so much so that its people
are weak, indolent, and apathetic; and "mountain-
eers," as they call the inhabitants of the hill-country,
are employed in considerable numbers to do the
heavy work, at extra wages. Perhaps it was from
the same reason that many of the houses on the great
estates, and the grounds about them, appeared neglected beyond what one would expect to see on such
valuable domains.

If I remember well, the soil along the road all the
way, or nearly all the way to St. Julien, showed little
variation, being mostly of coarse gravel—quite coarse,
the pebbles being as large as hickory and hazel nuts.

Nor could I notice much variety in the modes of training or in the degree of care bestowed. Westerly winds from the ocean often sweep violently over the peninsula, and, but for the very low training, would make havoc among high trellis or staked vines, and possibly I have here discovered a second reason for a custom to explain which a very high authority could give me no other than that it was to let the plow pass.

I noticed the shoots that had mounted above the tops of the laths of the trellis had a close cropped look, as if they had been trimmed like a hedge. And the driver said it was so; that it was usual at blossoming time to mow off, with a short scythe, both the tops and sides of the vines, in order to clear the way for the oxen and plow, which also served instead of pinching in.

LA TOUR.

The sight near at hand of "a stern round tower of other days" admonished me I was entering the domain of Château la Tour, one of the three reigning houses of Haut Médoc, by decree of the Bordeaux Chamber of Commerce and the suffrages of princely drinkers the world over ranking number one in a classification of a select sixty chosen from among the many thousand "*vignobles*" of a district where all is choice and fine. When the Franks invaded Gaul and first drank of the juice of its grapes, they honored the vine that bore them with the name *vigne noble* (noble vine), whence comes *vignoble*, the French for vineyard. But La Tour is more than noble. It has been crowned a king.

Surrounded by a field of little low vines, as insignificant to look at as any of the others, stood a handsome new chateau, with press-house, store-houses, stables, etc., close by, while apàrt from all, and rising from among the hop-o-my-thumb trellis was a stately antique tower, giving dignity, character, interest, and name to the place.

A gentleman of distinguished look, with two ladies, was walking toward the house as I drew near. I saluted him, and asked permission to walk about the property. "If Monsieur will be good enough to wait a moment, the '*regisseur*' will be here and will conduct him." The regisseur, or steward coming up, I was presented and turned over to him. He showed me to the press-house. A pile of grapes, already stemmed, was heaped in conical form on the pressoir, and five or six men, with trowsers rolled above the knees, were trotting about in a circle, trampling the pile under foot, beginning at the outer circumference,

and gradually contracting their circuit till they met
in the middle and on the top of the cone. This they
call "*fouler à pied*" (crushing with feet). There
might be a cleaner way of doing the thing; I don't
think there could be a fouler.

The regisseur made no apology for the sight, nor
did the trotters seem the least ashamed. Wherever
I went that day, except at the Averous farm and
Château Lafitte, this mode of crushing was in practice. It is said no other so effectually crushes the
pulp without breaking the seed—in fact, that it is
important for the quality of the wine that it be
trodden out with naked feet. It is also said, and
very truly, that soap and water will cleanse the feet
as well as the hands.

At one place I visited I inquired of the workmen
if they washed their feet before trampling on the
grapes, and was told they did not. One of them enlightened my ignorance by explaining that wine had
the power to fling off all impurities, so that it was of
no sort of consequence how free they made with it.
No doubt there is a good deal to be said on the other side of this question of dirt. I confess that what
I saw and heard disturbed my old notions. At all
events, the Médoc vintagers acted as if quite sure of
their chemical deductions, and would walk with bare

feet slap dash through puddle and mud, and mount the juicy heap with the assured tread of men firmly grounded in their principles.

Several of the vats at La Tour were already full, and fermentation was well under way in some of them, but none were covered with any thing at all; and this was the case at many other houses.

On my asking the regisseur how long he allowed the wine to remain in the vat, he told me the period varied from three or four days up to one month. This astonished me. He explained that, though fermentation might fully accomplish itself in a fortnight at farthest, even when the season was bad for ripening, yet that in such case it was needful to give time not only for fermentation to come to an end, but also for the greener portions of the pulp remaining undecomposed and suspended in the liquid to sink to the bottom. And that very vintage, he said, would need a month in the vat.

The regisseur's account of the precautions taken in drawing off, barreling, filling up, etc., satisfied me the great reputation of Médoc wines was much more the result of care and skill in "conducting" it, as they say, from the vat to the bottle, than is generally supposed.

When the time comes for drawing off from the

vats and putting into barrels, they proceed as follows: a sufficient number of new "barriques" (of nearly sixty gallons' capacity) to contain all the first quality of wine that has been made in all the vats are prepared by a simple washing with tepid water, followed by a rinsing with wine or brandy, and then drained until quite dry, and arranged in one or more rows on the floor of the cellar.

The vat is then tapped at the bottom, and the wine allowed to flow into a large tub at its foot, whence it is dipped out by means of oblong buckets, poured into the two-man tubs with sticks for handles before described, and carried and distributed among the barriques, not by entirely filling first one and then another, but by carefully dividing the contents of the vat equally among them all, so that when the man at the faucet, seeing the liquid begin to' run somewhat thick, turns the key and shuts off all farther flow, each and every barrique shall have received its equal share of the contents of the vat. The same process being repeated with each of the other vats, it follows that the wine in the barriques is uniform in color and quality, which is especially important where such small receptacles are used. Thus is made quality number one.

The turbid wine left in the vat is then drawn off, and set apart as quality number three.

The mass of skins and seeds still remaining at the bottom, and which is called "*rapé*," is then pressed with a machine resembling circular cider-presses, such as are now seen in use, and makes quality number four. Formerly numbers three and four were put together.

Number two is made from fruit grown on inferior soils or exposures, or that is, from any cause, imperfectly ripened. Grapes from young vines are also deemed unfit to mingle their juice with number one.

During the first month after the drawing off, the bungs are allowed to rest loosely on their holes, and twice a week the barriques are filled up. At the end of the month linen is wrapped about the bungs, and they are driven home. After that the filling up is done once a week. In March the wine is again drawn off. This is again done in June, as well as in October or November; but by some the June drawing off is omitted. In March of the second year the wine is again drawn off, after which the position of the barriques is so far changed, by turning them slightly on one side, that the bung shall always be wet, and the air-bubble rest a few inches away from it. From this time on, the drawing off is done only twice yearly, in March and August. At the end of three or four years the wine is ready for the bottle and for the market. In the drier climate of our country I am sure the term might be shortened by one third.

Great care is taken to keep the wine from any access of air when being drawn off. The common way is to place the empty cask beside the full one, connect the two with a tube of gutta-percha two feet long, allow the contents of the full one to flow into the other till the quantity is equal in both, apply a strong bellows to the bung-hole of the cask to be emptied, fitting it tightly, and blow out the remainder. Another plan is to place the full one immediately over the empty one, and let the contents flow into the latter through a tube reaching nearly to the bottom, in order that there shall be as little "churning" as possible.

The wastage from all causes between the first barreling and the final clarification for bottling is twenty-five per cent., and the cost of producing and conducting a gallon to the bottling stage is about seventy cents; with interest added, it would amount to a dollar. Considering that this estimate assumes the yield of an acre to be 250 gallons, which is about the average both of favorable and unfavorable soils,

it shows that labor and care are needed, as well as
silicious and ferruginous gravel, and that not wholly
to luck does Médoc owe its high reputation.

It seems the climate of Médoc is too damp to permit of storing wine in cellars under ground. Fortunately, however, and perhaps from the same cause,
it may be safely kept in store-rooms on the ground
level. During the first one or two years it is stored
in an ordinarily light but close apartment; after that
it is kept sedulously in the dark, as well as from all
access of outside air. It is deemed a sign that the
room is not dark enough if the mould which accu-
mulates largely on the casks is green instead of
white.

About ten feet above the floor of the store-room
is a ceiling which forms an attic overhead. This
attic is kept as close in all weathers as the store-room
itself, and a pretty warm place it must be in mid-
summer. The safety of the wine seems to depend
on keeping the temperature, whatever it may be, as
even as possible, since it is by changes of temperature that the wine in the cask is made to swell or diminish, thereby respiring new air, as it were.

Wine of the vintage of 1865 was uncommonly
good, and so was the regisseur to give me some of it.
He took me into the dark and musty inner apartment

where it lay, and there, on the very spot of its origin, I saw it drawn from the original package. Of course I found it the best wine, either red or white, I had ever tasted. Nevertheless, I was not dismayed, and I turned away from the precincts of La Tour with more hope and faith than ever in the Norton's Virginia Seedling.

PICHON-LONGUEVILLE.

From La Tour I was driven to the beautiful chateau of Pichon-Longueville, owned by a baron of that name. Looking at it, I wondered if the time would ever come for American vine-dressers to build houses like it from the profits of a hundred acres of ground too poor to bear mullens. Finding my way directly to the press-house, without troubling any one to give me permission or show me through, I was glad to find the workmen lounging about in the interval between dining and going to work again—the best time for getting questions answered. All was much the same as at La Tour, except that every vat in which fermentation had begun was covered with matched boards, closely fitted and plastered to the rim with clay or some kind of cement, so as to allow no escape for gas except through a tin tube of siphon shape, with its upper mouth submerged in a vessel

of water—an apparatus well known in this country.
Great importance is attached to the siphon and wa-
ter-dish at Pichon-Longueville, but why did I meet
with it nowhere else in Médoc?

Outside the gateway of the chateau was a compost-
heap, to which mud from the river marshes was being hauled, as well as stable manure. They told me
the compost thus formed and well rotted was the only
manure tolerated in Médoc, and that even this was
feared by some proprietors, who enriched their vineyards with mud alone, or turf, or swampy earth, at
the risk of debasing the natural soil, since large quantities are required if no richer material is mingled
with these. Compost is applied in various ways. At
Pichon-Longueville they spread it over the surface
and plow it in. Others fill with it little excavations
made around the feet of the vines, while others again
bury it in trenches midway between the rows, eight
inches wide, and deep enough to escape the plow.
The ill effects of manure on the quality of the wine
are not supposed to accrue in any direct manner, but
to result simply from the sap and luxuriance of the
plant which it induces. Freshly-manured vines and
those newly planted are placed in the same category,
the fruit of both being deemed equally unfit for
growing wine of the first quality; but, were not ma-

nuring sparingly and carefully done in Médoc, I am
sure the wine affected by it would go down at least
one step lower in the scale. There are proprietors
who do not manure their vines oftener than every
twenty years—as those of Léoville, for instance. The
greater number do it every seven, eight, nine, or ten
years, while some wait only five, and some only three.

LAFITTE.

After a pretty wide circuit, which brought within
view many celebrated estates, I made the next halt
at Château Lafitte. There, as at the Averous farm,
they did not dance upon the grapes, the stemming
process giving all the crushing thought necessary.
Now, through nearly every wine district of France,
they will tell you that crushing with bare feet is so
important, no considerations can be allowed to dispense with it. But do they not dispense with it at
Lafitte? and is not Lafitte a chateau of the first class?
Perhaps the omission of the ceremony there is an in-
novation of the present owner, Sir William Scott.
The British aristocracy are growing so fond of the
wines of Médoc, a good deal of its soil is getting to
be owned across the Channel; but, naturally enough,
they like to import as little of it in solution as can be
helped.

I saw them using the small round press I have mentioned to extract the juice of the few berries, mostly unripe, which still adhered to the stems after being rubbed.

A cursory observation would fail to detect in the pebbly surface at Lafitte, any more than at La Tour, any thing to distinguish it from a good many others growing wines of only second, third, fourth, or fifth class, or no class at all. But the uniformity is only apparent, and there is nothing occult in the matter. The ground of the Lafitte vineyards is of the following composition:

Silicious pebbles, nut size............	629.00 parts.
Fine sand.................................	283.00 "
Pure silex	62.20 "
Humus....................................	12.80 "
Alumina..................................	7.50 "
Lime	40.00 "
Iron.......................................	86.00 "
Loss......................................	4.50 "

Such is the composition of a soil capable of producing the very best wine. Next in excellence is a sandy surface underlaid with quite fine silicious gravel. After these two comes a surface of limestone pebbles, immediately resting on strong beds of shelly limestone or marly clay; and, last of all, soils where clay predominates.

Iron, forming nearly nine per cent. of the choice soil of Lafitte, is found in similar proportions in all other choice Médoc soils of the gravelly kind, and it is well known that it causes the wines grown upon such to deepen in color as they grow older, instead of fading, as is usual. The small proportion of lime, only four per cent., will also be noticed by those curious in grape soils, as also the general poverty of the whole mass.

Draining has long been practiced in the tenacious *alios*, or hard-pan subsoil of much of the Médoc region. Tile are already in use, yet many still insist that the old-fashioned brush-wood and broken stone drains are better.

LÉOVILLE.

The next domain I stopped at was nearly two miles from the last, in the adjoining commune of St. Julien. Château Léoville is of the second class. There, too, the vats had only loose boards for covering. Thus had I seen since morning vats wholly uncovered, vats covered with closely-sealed boards, others with loose boards, and others still with only a float of grape-stems. When such diversity of practice is found among skillful and practiced wine-makers, is it not best for the American beginner to buy some book that will instruct him in but one way of conducting the wine through every step of its progress, and confidingly follow all its teachings, without troubling himself with running after his own facts or spinning his own theories? I think not; and the method of this volume is to present—as it was of my explorations to collect—as many noteworthy facts as possible, let them perplex and puzzle as they may.

All the arrangements at Léoville were the best I had any where seen, and for the grapes they gave me I can pay them this compliment, that they were almost as good as I had eaten at home.

COS D'ESTOURNEL.

Another wide circuit, and I found myself driving by a carved stone gateway showing the royal arms of Britain as large as life. It was the entrance to Château Cos d'Estournel, owned by the heirs of an English gentleman named Martyn. Soon after passing it the driver drew up before a second chateau, belonging to the same family, called Pomys, and a fine old building too. A polite old Frenchman received me with an apology for the absence of the director, and showed me through not only the vines and winehouses, but the chateau, barns, stables, etc., very well worth seeing, and all bearing the stamp of English

order and neatness. But all his English associations
had failed to make an Englishman of my old conductor, or he would never have declined the money I
hesitatingly offered on taking leave.

The grating upon which they were stemming
grapes in the press-room of Pomys was framed of
oak bars one inch thick on the face and two and a
half inches deep, and the meshes, or openings, were
one inch wide by eighteen long. On the grating the
fruit was rubbed by means of a rake, also of oak, the
teeth being of the same stuff and dimensions as the
bars of the grating, set edgewise to the line of the
handle, and sharpened at the ends. The handle was
long.

In this connection I will describe a utensil for
stemming grapes which I think the best I have yet
seen. It is the one used at the Longworth Wine-
house in Cincinnati.

A tub flaring at the top, three feet high and four
feet across at its greatest diameter, is fitted with a
cover made of one-inch thick white-oak board, which
rests on shoulders that sustain it 28 inches above the
bottom, and seven inches below the top of the tub.
The cover has a strong cross-piece on the under side
to keep it from warping. It is pierced with holes of
the diameter of one inch on the upper surface, and

an inch and a half on the lower, the holes being four
inches apart, measuring from centre to centre.

Wine of Cos d'Estournel, which ranks in the second class, sells for 2500 francs per cask of four barriques, called "tonneau," while that of its next neighbor, Pomys, of no class at all, brings only 1000 francs.

The market value of the various classes will appear
in the following table, which gives the prices established for the vintage of 1862. Each tonneau containing 912 litres, and each litre being equal to
1.760773 pint imperial measure, and the franc being
equal to about 20 cents in real money, every reader
may reckon for himself.

1st class.............................4000 francs.
2d   " 3000   "
3d   " 2000   "
4th " 1800   "
5th " 1500   "
Bourgeois supérieure...................1400   "

The dinner to which I had been invited was given
in honor of the reunion of six college mates, of
whom two were the young Messrs. Averous, two were
young and very jolly priests, one was an Englishman, and the other a Bostonian. It was pleasant to
drink authentic Médoc in its very home, and equally
pleasant to witness the enjoyment of the college

friends. The priests were as good fellows as any of the rest, but French priests are never required to assume a vinegar aspect, and drink melted ice on principle.

CHAPTER V.

SAUTERNE.

I GOT upon the boat next day, and returned to Bordeaux once again, whence, on the following morning, I took the cars for Agen by a route that ascended the valley of the River Garonne. At Langon a gentleman entered the carriage where I was whom I found could give me full information concerning the vine district I had just traversed in the preceding half hour's ride. It was the district where is grown the fine white wines known under the general name of Sauterne.

He told me the soil was in some places of gravel, in others of sand, and in others of clay mixed with sand and underlaid with limestone. They plant their vines about three feet apart in both directions. They prune them low, the two or three canes allowed on each stock, or souche, being cut back to two eyes each. Leaf-pruning is practised to excess. Beginning early in September, they proceed with it gradu-

ally, but severely, so that before vintage, the fruit,
little by little robbed of all its natural shelter, hangs
naked to the sun's rays.

Three gatherings are made; the first culls from
each bunch only a few excessively ripe berries, the
second takes such as have ripened to the same excessive degree since the first, and the third sweeps in
all the remainder. This protracted vintage hardly
comes to an end before November. Wine of the
first gathering is called "head wine;" of the second,
"middle wine;" and of the third, "tail wine." About
ten per cent. of the whole is of the extremely precious first fruit, and about forty per cent. is of the
second quality. The average yield is about the same
as in Médoc.

They manure once in five years, plow thoroughly,
and cultivate by hand as well, and, from what I
learned of my railway-carriage informant and from
other sources, are even more exact and careful in
the conduct of their delicate wines than those I had
just left.

After the first two or three years Sauterne wine
is transferred from the barriques into very large tuns
called "foudres," which hold nearly 2500 gallons.
While remaining in barriques, it is drawn off three
times a year, and filled up twice a week. In foudres

it is drawn off twice a year, and filled up once a week.

Haut-Sauterne is of marvelous delicacy, and is the French rival of the white wines of the Rhinegau.

At half past eight next morning the landlord of the hotel at Agen said, "Certainly, if the gentleman wishes it, I will wake up the cook; but the gentleman will enjoy his breakfast better if he will wait till eleven o'clock, when we have our table d'hôte." "I will wait, certainly," I replied, too polite by far; and it is true I enjoyed my breakfast when I got it, as any fool might who had waited two hours and a half. The table was lined with those knights-errant of modern times, as Irving calls them, commercial travelers, who overrun all civilized countries, beating the reveille for customers. Of a truth drumming is a hungry exercise, or else Frenchmen are better eaters than I am used to see; and I here note that Americans can not compare with Frenchmen at trencher-play, either for quantity or rapidity. The commercial eaters at Agen waited patiently, it is true, while each of the eight or ten courses was being served, but, once started, their speed from station to station could not be emulated by us.

CHAPTER VI.

LANGUEDOC.

FROM Agen to Toulouse, and thence to Beziers,
I took a third-class car. The seats were hard,
with hard backs. It rained hard without, and they
smoked hard within. The stoppages were frequent,
and the speed under twelve miles an hour. It was
a hard journey; and I would warn all travelers on
the Continent who may like to study French character, and sound the feelings of the working classes
throughout France, that they had better take good
lodgings in Paris, read carefully some book on the
subject written by an unprejudiced Englishman and
rely on its statements, instead of going bumping
about among blue blouses and wooden shoes, as I did.

After three or four hours, the rolling of the r-r-r's
of the incoming passengers reminded me we were
entering the borders of ancient Languedoc, so named
from the old language of its people, which was not
French, but the *langue d'oc*—a country of great tra-

ditions—Roman and civilized, rich and enlightened, when Paris was but a seat of barbarians—Paris, which now scorns as provincial the rich roll of the Languedocian tongue.

The hands and feet of the peasantry of Languedoc are so small, it is difficult to believe them used to hard work. They are of a superior bodily constitution—blooded animals of Greek and Roman descent; and centuries of labor, deprivation, oppression, persecution, and ravages of war have failed to bend their backs, hollow their chests, make their feet flat, their hands broad, or their waists large. Their eyes are large and black, and their teeth white and clean.

Having left behind me the Médoc and Sauterne districts, where, as is the boast of their cultivators, "Nature does much, but man does more," I was approaching those happier southern regions where Nature does a good deal more, and leaves man not near so much to do; where the kindly sun, knowing how languid the ardor of his rays renders the arm of man, takes upon himself the greater task, and performs it by a fuller efflux of creative power, leaving only the lesser one for poor lazy mortals to do.

Happy, beautiful, fruitful Languedoc! I will one day see you again, please God.

So frequent were the stoppages and changes of

passengers, that in the course of the day a good many different people successively occupied the benches neighboring to mine. With most of them I managed to have some conversation, and they had a good deal to say on the subject of vine-culture, but it all related to vines trained "en souche-basse," which, as I have said, may be translated "stump," "stock," or "stool." As I looked out through the windows of the carriage, it meant beautiful bushes, flourishing over hill, hollow, and plain, from rail-track to horizon, as good to see as corn on a prairie. Not one brown stake or lath was there to mar the green array. Nature, in her strength, had flung the crutch away.

But I was not allowed to do all the questioning, and, after receiving my fair share of information, was required to give some in return.

"The gentleman is from America," I heard one of my companions say to a woman sitting next him. "Ask him, then," she said, "if the men there can have as many wives as they like."

"Yes, madam; some of us take one, two, or as many more as we can support, but we do it to carry out our conscientious convictions—just as your monks, from an equally high principle, refuse to have even one wife."

D

"Horrible!" she cried.

"Abominable!" said a priest who had just then turned round to listen.

"Not a bit abominable," growled a gray old farmer from behind me. "Much better have too many wives than none at all."

The priest smiled and took snuff. Neither he nor I had ever before heard the case put in just that form. The woman looked down and said nothing.

Beziers, where I arrived at evening, is one of the principal cities of the Department of L'Herault, which produces more wine than any other in France. The city is in view of the Mediterranean, has thirty thousand inhabitants, and is neither very neat, orderly, nor beautiful. In going to Beziers I had plunged into the very middle of things, but, as at Pauillac, I had no clew to get at them—not so much as the name of a single soul in all the place.

After supper I went to the theatre, and there had the good fortune to sit beside a gentleman with whom I soon got acquainted, and in whom I found the very person I wanted. He gave me his card at parting, and invited me to visit him the next day at his country residence in the village of Boujon, some six miles from the city. I found M. L—— at his distillery, adjoining his cellars. The stills were of

the newest and most complex fashion, more expeditious than the old well-known copper affairs, such as,
in our country, good whisky is made with, and made
probably no exception to the old rule regarding all
things good to eat or drink, that, to arrive at excellence, we must travel in a slow coach. The liquor
made at M. L——'s is called "*trois-six*"—three-six
literally, but why, I forgot to ask or forget to remember. It is brandy, of course; but, for some good
reason, what they make in Languedoc, and in other
places too, is not called by that name. Some of the
wine they were distilling when I was there they offered me to taste. It was the five-cent quality of
which I have spoken—not five cents per litre, but per
gallon.

In the press-house they showed me the vats, which
were great square cisterns of cement, and covered
loosely with boards. Cement is admitted to be inferior to wood, but is cheaper. The cellar of M.
L—— was well arranged, well kept, and very large,
there being ten great casks, each of the capacity of
about ten thousand gallons. All were not filled
with five-cent wine, however, some of them containing what was considerably better, and not intended
for distillation.

As we walked in the vineyards, I could see the

soil was mostly of good gravelly loam, and, if it was devoted to growing such extremely cheap wine as I have named, it was not because it could not produce good crops of wheat. When M. L—— mentioned that the vines of Languedoc were mostly of Spanish varieties, introduced when that province was under the same crown with Aragon, I recalled that those of California, and which are trained in the same way, and produce wine so different from what is grown any where else in our country, were also of Spanish origin.

M. L—— took me to his house, a very snug bachelor's home, where, cosily seated by the fire, which damp weather and the approach of night rendered comfortable, was another gentleman, who had come from the city to be his company during vintage time. They had been college chums together half a century before. Now, considering that at the last vintage gathering I had attended all the company were college classmates, perhaps I may safely state in this place that it is a French custom for college friends to meet at vintage instead of Christmas. With the two old friends I tasted several kinds of the older and better wines of the neighborhood; then, declining farther hospitality, I returned to Beziers.

I resolved I would go farther into the subject of

growing wine on the Languedoc plan. But to do this as thoroughly as I ought would require weeks of delay. I must, for the present, turn away from the shores of the Mediterranean, and journey northward and eastward into the great red-wine country of Burgundy, if I would catch some of the fleeting vintage hours before the harvest-girls have gathered the latest, ripest clusters from the *"Côte d'or"*—the hill-side of gold. The same necessity obliged me to travel in the night-time, so that I reached Lyons without seeing much of the important valley of the Rhone.

The same day I pushed on to Macon, where I staid overnight, and the following morning went on to Beaune, a small town at the foot of the Côte d'or, and not very far from Dijon, the old capital of Burgundy.

CHAPTER VII.

BURGUNDY AND THE CÔTE D'OR.

OF a verity, the heritage of the old dukes was a goodly one. To the traveler who goes through this fair and rich province, the wonder is, that, with such wide and fertile valleys to nourish them—such strong, full-bodied drink to nerve them, such rivers at their feet and mountains at their backs, the rulers of Burgundy did not become kings of France.

I had never tasted a drop of authentic Burgundy wine in all my life. Few people have who live across the seas, for it does not bear transportation, notwithstanding its alcoholic strength, which exceeds that of Bordeaux wine. Its market being, from the cause just given, a limited one, it is sold cheaper than Bordeaux of equal quality.

Being a stranger, I made no useless attempt to obtain a very choice sample, but called for a bottle of such as bore a moderate price, and, being fatigued, went so far as to drink a tumbler full of it. It was very palatable and refreshing.

I had never drunk any before.

I have never drunk any since.

I shall never drink any more.

Presently I will tell you why.

At my special request, the hotel-keeper procured for me a driver who had been a vine-dresser, and was well informed concerning what I wanted to know about. I think it important the drivers of public carriages should be men of intelligence, for a deal of information is sought of them. They hold, in fact, the not very mean position of instructors to the traveling world. In Paris some seven hundred ex-priests sit upon the box and hold the reins. Maybe it is from having educated tutors such as these that our American young gentlemen so easily learn all the ways of that wonderful city.

Telling the driver what I wanted, and seating myself beside him, I was driven first toward the village of Bligny. The vigorous growth of the vines, as well as the appearance of the soil, showed the great plain across which our course lay to be a rich one. Formerly, my guide said, there were but few vineyards on the plain, but of late years they threatened to crowd out every thing else, owing to the increased demand for wine of the quality there grown.

The souches observed no order whatever. Though at first planting they are set in regular lines five feet apart, and eighteen inches distant from each other within the lines, the system of layering, to supply losses or restore decaying vines, causes them soon to break ranks and straggle hither and yon most confusedly. One result of layering is, the cultivation must be done by hand. They cultivate thoroughly three times a year, once in March, once before blossoming, and again after the grapes are well formed. I nowhere saw any trellis, but learned they were slowly and doubtingly being introduced. They involve a reconstruction of the entire vineyard where adopted, as well as a radical change in the system of training. Some of the vines appeared to have been pruned very close, leaving only one or two eyes to each cane, but reserving several canes. These were varieties whose habit is to bear the fruit close to the souche, or old stock. Others, with a tendency to bear from buds farther out on the cane, had three and four eyes.

I inquired what was the price obtained for such wine as was commonly produced on the plain, and learned it usually brought 65 francs the " piece," containing 228 litres, the same as the Bordeaux barrique, but, owing to the poorness of the present year's crop, resulting from hail as well as excessive rains, 120 francs was being demanded.

The ordinary qualities of Burgundy wine, as, indeed, of all other wines grown in France, are disposed of early, and are generally consumed within one, two, or three years, while fine qualities must be kept and cared for during from three to six years.   My guide, mentioning this, said, for his part, he would rather own a vineyard on the plain than on the Côte d'or.

On the plains all kinds of manure seem to be used: stable-sweepings, oil-cake, bone-dust, guano, and rags. I believe, though, manure is only applied in layering, that operation being so frequent that the whole vineyard gets enriched by the share allotted to the layered plants.   The usual number of souches to an acre is five thousand.

The vintage had nearly come to an end; only here and there, at wide intervals, did we encounter the bands at their work, or a wagon on its way to the press-house.

"What wages do farm laborers get?" I asked.

"Three francs a day; how much do you pay in America?"

"A little over five; but, aside from food, it will not purchase as much there as three will here."

I hope it made the victim of imperial tyranny more contented with his lot to learn this.

Arrived at Bligny, I was set down at a large wine-

D 2

house where they were pressing white wine with a
press of the old fashion, having a lever which is
worked with a screw. They showed me into the
"cave," as underground cellars are called, where the
must was fermenting in barriques. All were over-
flowing in froth and impurities at the bung-holes,
it being the practice to insure this effect by completely filling the casks instead of leaving a void, as
we do.

. The fruit was being crushed in a grape-mill. The
juice running without pressure was to be set aside,
and afterward mixed with a third or fourth part of
the expressed must. The workmen at Bligny stoutly
defended their old lever and screw against all new
comers in the shape of patent presses, of which Burgundy is full. Its advantage seemed to lie in the
spring of the wood, but this might easily be obtained
in some less cumbersome way. In the course of the
day I saw several presses of late invention better
than any I had seen at home.

Where I next stopped red wine was being made.
They were stemming the fruit by rubbing it on
small basket-work sieves or gratings, slightly bag-
ging in the middle, resting on tubs into which the
crushed berries fell. The work was quickly done.

The vats I found to be much smaller than those

used in Médoc, and with no covers at all. A smell of vinegar came from the "châpeau," as the floating mass of skin and seeds which rises to the top is named, while swarms of little gnats, always a bad sign, hovered above it. The foreman told me that before putting the "rape" to press the top of the châpeau would be carefully pared off down to where it smelled as it should. As a precaution against acetous fermentation, they sometimes sift over the top of the châpeau a coating of plaster of Paris the third of an inch thick. I don't think, however, this would be done in making fine wines, but for an ordinary quality should think it good.

When I inquired how long the wine would remain in the vat, they told me two weeks; but said that last year, which was as good for ripening as the present one was bad, only four or five days was found to be enough.

One half the wine-makers in France stem the grapes before putting them to ferment, and the other half do not. In Burgundy it is not done except when the ripening has been imperfect. The reasons given for this by those Burgundians of whom I inquired did not appear to me entirely reasonable, but doubtless the practice is founded in reason for all that. People who inherit wise customs born of the experi-

ence of a remote ancestry do not always find themselves in possession of the rationale of those customs.

Among the consequences of keeping the stems out
of the vat would be a disease called "bitter," common in French wines, and a flat taste; the others
which were named I have forgotten. One person
told me the corrective virtue of the stems consisted
in their acidity, and another thought it lay in the
tannin. There is hardly any tannin in stems, and as
for acidity, it might be better obtained from unripe
grapes, as is done in Médoc, where juice of the folle
blanche is mixed with that of the malbec, to correct
the flatness of the latter.

Grape-mills are coming into fashion in Burgundy,
but crushing with feet is still the general practice.
It is usual to keep red wine above ground until March
following the vintage, and after that in "caves,"
which are most commonly arched. There is no mystery in the fact that wine will keep above ground in
the intense summer heat of Languedoc, while in the
cold climate of Burgundy it must go below or spoil,
for the wines of the south are remarkably free of
acids.

At the "Hospice," the property of an endowed
hospital in Beaune, I found the arrangements excellent. The foreman, or one who seemed such, ex-

plained things like one who understood what he was
talking about.   The vats, which are constructed like
those of Médoc, only lower and wider in proportion
to their height, hold about a thousand gallons each.
They usually have no covers, but sometimes a false
top is adjusted below the level of the surface of the
must to keep the châpeau always submerged.   Often
it is necessary to resort to artificial heat in aid of the
fermentation.   One way is to heat a portion of the
must to the boiling point and then return it to the
vat.   A good temperature for the must to show when
a thermometer is introduced is from 80° to 90° Fahrenheit.   The experiment has been tried of adding
a quantity of white wine, itself in an active state of
fermentation, to serve as a kind of yeast.

So long as things work well in the vats, nothing of
the kind is needed.   But there is another mode of
rousing up the slackening process, and at the same
time bringing the skins and seeds which have settled
to the bottom into contact with the new-made alcohol, so that the latter may combine well with the coloring matter they contain.   This consists in stirring
up the whole mass from bottom to top.   It is done
twice during the process of fermentation.   It needs
a good one hour's hard work each time.   It is done
by men.   It takes four men to do it well.   They all

strip naked—naked as Adam when he was good—
and then they go in—into the wine-vat—chin-deep
they go in, and there, with feet and hands, fingers
and toes, turn over, stir about, and mix the liquid
that was getting clear with the pomace that was depositing itself, and

> " Make the gruel thick and slab,
> And like a hell-broth boil and bubble."

The nice, sweet Bordelais man only puts his foot
in it, but the Burgundian goes the whole figure.

It is done to give the wine a full body.

They call it fermenting on the skin.

He who explained all this to my astonished mind
avowed it with the simple frankness of a Feejee
cannibal who admits his fondness for what he calls
" long pork." But the Feejee people are only hea-
thens.

In Lamartine's letter written to justify the Emperor's expedition to Mexico, to set up an empire
there which should hold the American Union in
check—in which letter the author earned for himself a good pension — rests his case on the sole
ground that our peoples' manners are bad. It is
true, our manners are bad, and maybe Napoleon
did right to punish us for them as he did. Certainly we can not dance as well as Frenchmen; but

oh, Lamartine, owner of many vineyards! can worse dancing be done than in a vat·of wine? or worse manners possibly be than afterward offering it to be drunk?

At the Hospice I first heard of this strange custom, but repeated inquiry afterward confirmed the story. Nor is the custom confined to Burgundy alone, or to France alone. "Once," say they, "our wines fermented on the skin only one or two days, and were light in color and taste; but the consumers of late years demand a deeper color and richer taste, so in we go."

Stirring up with poles they tried, but the warmth of the human body was wanting, and the result, they say, was not good. Besides, it was hard work.

To prove, however, that no good reason exists for the practice, the Vicomte de Vergnette Lamotte tells us he succeeds perfectly in obtaining the deepest color, and even more alcohol than fermentation in open vats can give, without stirring up (*fouler*) of any sort, simply by using a large cask, with an opening twelve inches by eight at the place for the bung. In such a vessel he has allowed the must to work for twenty-two days—quite beyond any period that would be safe with an open vat in Burgundy.

But does this amount to any thing more than par-

tially covering the common open-mouthed vat? For
the benefit, however, of any who may choose to try
the Vicomte's plan, I will add that the lower door,
or man-hole, through which the marc or "rape" must
be withdrawn after the wine has run off, is fitted on
the outside, and held with two bolts. Closing from
the inside, it could not be opened, owing to the pressure of the marc against it.

From 60° to 70° Fahrenheit is thought a favorable
temperature for the fermenting-room. Doctor Gall,
the German writer, thinks differently, and recom-
mends to heat up gradually, by means of a stove, to
about 80°, and keep it so. I think I have heard
stoves were sometimes used in the fermenting-rooms
in Champagne.

A good proportion to observe in the form of open
vats is one that will give the liquid the same depth
as diameter. New casks are preferred in Burgundy
as well as in Médoc. Pains are taken to keep them
full, and they draw off frequently.

I was disappointed to find, on leaving the Hospice, that I had so badly reckoned the time a visit to
the celebrated "Clos Vougeot" had become impossible. Those travelers in Europe who visit sites of
ancient monasteries may observe that by a special
providence, as it were, they are usually found in the

midst of the most fertile meadows, the fattest grain-
fields, the richest fisheries, and most golden hill-sides.
One of these last, "*Clos Vougeot*" by name, found
itself included among the broad possessions of a
house of holy men—having been moved there, doubtless, by their meritorious faith—and for many long,
tranquil centuries its nectarean flow refreshed the
piety of an uninterrupted succession of jovial saints,
and elevated their souls almost to the ecstasy of gods
—heathen gods I mean.

So long as those good men possessed the Clos Vougeot, no impure admixture was allowed to taint its
virginal soil, nor was any layering done, or other
means practiced to force the yield of its old patri-
archal vines, which were allowed to attain the incred-
ible age of four or five hundred years, and the whole
plantation of eighty acres required to give only about
twelve hundred gallons yearly.

But the French Revolution came, and Jacobins,
Republicans, and sinners drove out the monks and
usurped their domain. Now see what followed.
The good wine soon ceased to flow for the impious
dispossessors, on whose lips the grapes, playing the
old trick of the apples of Sodom, turned to sour cider
instead. Was it that a miracle of divine wrath had
blighted the soil and its fruit, to punish the sacrilege?

No, only this: the secular owners had rooted up all the old growth, and planted new vines, which responded to their avaricious exactions with a yearly yield measuring eighteen thousand gallons in *quantity*—but in *quality*, oh how inferior! It sells at present for about the price of fifth-class Médoc. Yet *Clos Vougeot* is the king of the Côte d'or.

Within limits, a law of Nature ordains that fine things shall not come in gross bulk. Diamonds and emeralds are not found in massive beds, like granite and limestone. Sable and ermine furs do not grow on the backs of buffaloes, neither can a lioness, however she may try, compete with a rabbit in the business of reproduction; and whoever hopes that a given vine-plant will bear one or two thousand gallons of choice wine to the acre, hopes against law—and hopes in vain. 150 gallons for the mean yield of the choicest and best; 250 for the mean yield of the totality of Médoc, the Côte d'or, and the Rhinegau; and for what comes after them, 500, 1000, 2000, and even 3000, very much according to quality—quantity according to quality, and quality according to quantity—this is the law as applied to the subject in hand.

The last portion of my drive that day was along the Côte d'or, and among vineyards of the delicate,

slender "pinot" variety. These, like those of the
plain, were set without regularity, and so closely that
it took 9000 of them to cover an acre. On the hills
they prune to only one cane, and on that one allow
from two to four eyes, and this notwithstanding the
disposition of the pinot to carry its fruit on the upper
buds. The souches were rather high, little pains being taken to keep them low. But there is a reason
for this: the fruit grown on a high souche is better,
for the same reason, probably, that on old vines is
better—namely, because the sap grows richer by
mounting slowly through hard and twisted stocks of
old wood. To leave, as we do in America, eight,
ten, or even twelve eyes to a cane, would be thought
murderous treatment in Burgundy, insuring a speedy
end to the victims; and, indeed, I don't know in
what part of France they would not think so. It is
to be seriously considered whether our plan of long
canes, bent in circles or bows, is not in the end ruinous—has not, in fact, ruined many a vineyard. True,
M. Guyot, of whose system so much has lately been
said in France, claims to have obtained great results
by leaving nine or ten buds on one cane, which cane
he extends along and close to the ground, and ties,
at the end, to a peg. But among the many objections brought against it is this, that, despite the free

use of manure which he advises, the vines yield poor
fruit, and soon wear out. And, apropos to our own
mode of bow-training, one of his opponents cites a
case where a vineyard was ruined by that very method, which, he declares, is nothing else than Guyot's,
only the cane is bent in one case, and kept horizontal
in the other, the effect on the circulation being in
both very much the same. Leaving on a given
souche three or four canes, each of them trimmed to
two or three eyes, is not considered by any means
bad practice with prolific varieties, on rich ground;
yet to select a single cane, and leave on it as many
as eight eyes, would be. The reason is, eyes remote
from the souche will generally bear more than those
near it. The habits of plants differ in this respect.
Some bear more fully on the lower eyes, and with
them long pruning would of course be safer.

The pinots on the poverty-stricken flanks of the
Côte d'or hardly looked able to yield an average of
150 gallons to the acre, which is, I think, the mean
product of the more celebrated vineyards there,
though, as in Médoc, 250 is the general average of
the whole hill. The Côte has a varied soil and subsoil. On low hills at the base, resting on alluvion, and
formed into elevations by the washing out of ravines
and other accidental depressions, the composition is

clayey, but with large portions of lime and iron. On friable magnesian limestone another soil is formed, unmistakably reddened with a large admixture of iron. The harder oolitic ledges, outcropping along the hill and sustaining a broad bench of gentle slope, give a surface soil containing iron and silex in large quantities. Another kind has a subsoil of limey marl. All of these yield the very highest qualities of wine. Here is an analysis of two kinds of soil and subsoil as made by a distinguished chemist:

Large and small gravel of limey nature.........	30.10	29.15
Carbonate of lime...............................	12.95	17.20
Carbonate of magnesia...........................	3.98	
Oxide of iron...................................	12.72	10.50
Alumina..	5.93	7.17
Silica...	28.93	32.98
Organic substances.............................	5.39	3.00
	100.00	100.00
While the corresponding subsoils showed		
Carbonate of lime...............................	88.00	78.00
Argillaceous substances.........................	12.00	22.00
	100.00	100.00

It will be noticed that one of these analyses shows about thirteen per cent. of iron, and the other between ten and eleven. The Lafitte soil in Médoc contained, it will be remembered, between eight and

nine per cent. Perhaps we have yet to learn how
important to the production of fine qualities of wine
is the presence of iron.

On estates growing fine wines, they apply manure
only when necessary to save the life of the vines; but
they periodically haul from the bottom of the hill
and restore to the soil its loss from washing, and the
effect of this is said to be remarkable.

Though the soil of the hill seemed to be of an im-
permeable nature, I could not learn that the ground
was ever dug up very deeply. To prepare it for
planting, they dig, along the slope and following the
course of the hill, trenches fourteen inches deep and
twelve inches wide. Crosswise on the bottom of a
trench the rooted plant is laid, with its top resting
for support against one of the sides. It is covered
with six inches of earth well pressed down. The top
is made to rise above the soil to the same height as
in the nursery. At the end of the winter, or, if the
planting is in spring, then in a fortnight after planting, the side of the trench against which the plant
leaned is pared away, so that the bank which served
in winter to shelter from the cold shall not any longer exclude the sunshine.

And this seems to be the only preparation the soil
receives for a new plantation. I had a vineyard

once planted in this way by a Burgundian, and the vines took root uncommonly well, forcing their way downward and sidewise into the tough clay subsoil far deeper than they could have done if planted in the common mode on ground trenched ever so deep. The trenches were kept free of weeds, and only gradually filled up. They were not entirely filled until the end of two years, or maybe three.

This is not very expensive, costing per acre, by contract, 100 francs, which represent thirty - three and a third days' work if the summer wages of three francs are allowed; but, as it is probably done at a time when wages are lower, we may call the labor-cost fifty days. The cost of manure and of the plants or cuttings are omitted.

The outlay of labor for cultivation is equally moderate—surprisingly so if we consider that 5000 or 9000 staked vines, standing in confounded confusion, are to be hoed by hand thrice in a season.

It is usually done by contract, for sixty dollars the hectare, or about twenty-four dollars per acre, representing forty days' labor. This covers all the work but harvesting, and includes laying down some 560 "*provins*," as they are called, which is the average yearly number of vines to be layered.

Why is this done?

To reinvigorate sickly plants or replace dead ones.

But why do as many as eleven per cent. get sick or die every year?

Because, where they are crowded together at the rate of 5000 or 9000 to the acre, they are suffocated and starved.

But why are they set so closely as to suffocate and starve each other?

In order to improve the quality and hasten the ripening of the grapes—wood and foliage being sacrificed to product.

I will recur to this system of "*provignage*" when I come to the vine-culture of Champagne, and describe how it is practiced. It does not necessarily exclude trellis-training or cultivating with a plow.

Burgundy has as yet known but little of the oïdium; the Côte d'or has absolutely escaped. Champagne is almost equally fortunate. Some have said the climates of those districts were too cold for the pestilent parasite to live there, but I found it in the Rhine vineyards, and it thrives in those of the Bordelais districts, quite as cold, I think, as the others.

It is known that young vines seldom have the disease. Now the system of "*provignage*" common to both Burgundy and Champagne rejuvenates the plants and keeps them always young, and on that

:

very account has been objected to as tending to deteriorate the quality of the crop, since good wine requires old vines to produce it.

Thus it may be that provignage keeps off the oïdium by keeping the vines always in a state of infancy.

I well remember a vine-dresser from Champagne, who, having purchased a decayed and rot-ravaged hill-side of Catawbas, near Cincinnati, about the year 1856, layered them all, and for years afterward continued to gather good crops, while all around him were being ruined by the scourge.

It is in view of the possibility that layering may be found a sufficient remedy for oïdium, as well as a means of restoring vines made sickly by its repeated attacks, that I have given the cost of cultivating by hand on the Burgundy plan.

I have said that in former times the Burgundians let the must ferment on the skins but one or two days, which gave only a light tint to the wine. They do the same, I understand, in Missouri, and the result is a white wine, properly so called, pinkish in tint, but not, for that reason, correctly termed red. I am sorry to learn that the Germans of Herman, who first taught me the value of the Norton's Virginia Seedling, and from whom we obtained roots to plant

E

the two first Norton vineyards in the Ohio Valley, should thus abuse its noble fruit. Mr. Hussman, in his book lately published, tells us they do it because a complete fermentation would render the wine too astringent. If this be so in his state, we must deny his claim to the plant, as a Missourian by adoption, based on the assumption that it succeeds better there than with us of this valley, for we have made a veritable fourteen-day red wine from it that none can say has any harsh quality, and which was received with respect by good tasters in France, and spoken of with praise in their first journal of viticulture.

Red wine and white differ materially, and in essential respects.

The Vicount de Vergnette Lamotte says, "White wines of the same year and of similar growth exhibit from the beginning a perfect identity; red wines, under analogous circumstances, often show very distinctly-marked differences.

"White wines, and wines from red grapes, but which have not fermented on the skin, are not subject to the same disease as true red wines, or yellow wines made of white grapes fermented on the skin.

"White wines are richer in alcohol and in acid salts than red wines of analagous growths. These last, on the other hand, contain a stronger proportion of tannin and extractive matter."

Doctor Guyot, at the same time physician, wine-grower, and author, says :

"White wines are generally diffusive stimulants of the nervous system; if they are light, they act rapidly on the organization, whereof they exalt all the functions. It seems they escape just as rapidly by the excreting organs of the skin and mucous surface, especially by the urinary ways; their action is, then, of short duration.

"On the contrary, red wines are tonic, and continuing stimulants of the nerves, the muscles, and digestive functions; their organic action, being slower, continues longer; they do not increase the perspiration nor the excretions, and their general action is astringent, persistent, and concentrated."

Doctor Ludwig Gall, physician and chemist, says:

"The greater amount of tannin in red wines fermented together with stalks, skins, and seeds, or even skins and seeds alone, seems to be the reason why they are generally preferred as a common beverage in Southern wine-growing countries to the white wines containing a greater amount of tartar.

"The effect of the high temperature of those countries in relaxing the muscles would become greater by the frequent use of a beverage containing much tartar and of a laxative character, while tannin tends to produce a greater contraction of the muscular system than any other substance in daily use.

"The Northern man, on the contrary, whose tension of muscles is naturally much greater, requires in his drink something that quickens his blood and promotes its circulation, rather than an astringent; and this is done by the alcohol in its diluted state, such as is found in good wines."

For the purposes of this last paragraph, ninety Americans in a hundred are of Southern constitution, and need a tonic rather than a stimulant.

A Parisian physician, prescribing for a delicate American patient, will nine times in ten order red wine. "I am cured of my dyspepsy," said one of these to me. "Did the red wine do it?" I asked. "No; I think it is the variety of courses at the table d'hôte. I think I shall give up the wine, being opposed to it on principle." Poor, inconsequential teetotaler! He could believe in the digestibility of soup, salmon, radishes, fried beans, cutlets, salad and chicken, cauliflower fricassee, salmi, blanquette, cheese, custard, pudding, tarts, syllabubs, raisins and almonds, cucumbers and melons, all jumbled into one meal, rather than in so simple a thing as red wine.

That self-devoted apostle and missionary, Professor Babrius, of Bordeaux, discoursing in 1840 on the influence of wine on civilization, speaks of the effect of French wines (red, of course) on the French people thus: "So long as wine was honored by all classes, the French people remained, in virtue of their brilliant qualities, the first of modern peoples. Courage, loyal and generous, gayety and vivacity of mind, patriotism, eloquence, an exquisite sentiment of personal dignity joined to an excessive politeness, an irresistible longing for a sweet sociability, were the principal traits of their character. When coffee, tea, and tobacco successively took their place among our habitudes, each of these agents, more or less deleterious, impressed a sensible alteration on this beautiful assemblage of distinguished traits."

And there was a good deal of ground for this self-laudation. The deep-thinking Babrius goes on to say:

"What distinguishes wine from all other drinks is its general action on the bodily economy. In moderate quantities it increases the energy of all the faculties. The heart, the brain, the organs of secretion, the muscular system, each acquire by its use a sensible augmentation of vitality.

"Wine acts generously on all our functions, fortifying and exciting them in harmony with each other, while other liquors act like those medicaments which expend their force on a single organ merely. Far from increasing the harmony of the system, their action can not fail to trouble it.

"Coffee, like wine, excites the vitality, but it stimulates only those portions of the brain which are the seat of the intellect, properly so called, and the speech. Its special property is to cause a flow of language, clear, lively, and facile, that is never troubled by the emotions of a warm conviction. Under the action of coffee the heart remains perfectly calm. It is the drinkers of coffee who have said you must not feel a sentiment if you would express it well. The decoction of coffee is the liquor of men of the world; it is the provocative to counterfeits of the truth, to cold and piercing sallies of wit, to specious argumentation—in fact, to all which makes up the charm of the elegant and blasé world of the saloons.

"Tea addresses itself directly neither to the heart nor the head. Its stimulation goes to the glands of the abdomen. The liver and reins respond strongly to its action. This explains why tea facilitates the digestion of indolent stomachs, and why its drinkers are inclined to moods of melancholy. They are cold, and talk little. Tea impresses on individuals and on nations which use it a slight tinge of hypochondria."

But as tea and coffee still hold only a subordinate place on the tables of the French, Babrius thinks that, though they may alter and distort French civilization, and turn it aside from its true end, it will not perish from their influence. What he fears will enervate, confuse, corrupt, and finally abolish us all, is tobacco.

The American people is in want of a drink. A nation has transplanted itself, but not its vines, from one hemisphere to another, and is thirsty. It is as important what we drink should be adapted to our climate, our temperament and institutions, as it is we should hold correct opinions on this, that, and the other subject. In fine, the liquor to mix daily in our blood, to act on our nerves, nourish our tissues, and qualify the vitality of every part of us, will control our destiny as much, at least, as what we learn in schools, read in newspapers, or hear from pulpits.

What shall we drink?

It will not answer in these days, with the deplorable results we have before us of the evils of water-drinking on one hand, and the evils of spirit-drinking on the other, to point to the springs and brooks, rivers and lakes, saying, "Share with the frogs and fishes, and four-footed beasts, the abundant washings of the earth's surface; there is enough for all."

We live in a dry climate, and under moral conditions exciting and exhausting to body, brain, and nerve.   That climate and those conditions have already, in the absence of any proper corrective, created a national temperament that responds with excessive sensibility to every exciting cause.   The pale, bony woman, who paralyzes her insides with unstinted draughts of liquid ice, and the restless, nervous man who consumes his with draughts equally unstinted of liquid fire, are types alike of our wretched condition as a people.   Dilution will not save us. Says my scientific friend, Doctor ——, "A low dew-point (dry air) and Republican institutions are inconsistent with the long duration of our race."

Now we don't want to pull down Republican institutions, nor can we raise up the too low dew-point. We must raise red wine, then; and this can be done, I will endeavor to prove, as easily and cheaply as in Burgundy, where it is to be had of good quality for four, five, and six cents a bottle.

Taken in the quantity of a quart daily for every adult, and a pint daily for each child, we may expect the following effects: It will slightly stupefy, and thereby soothe and quiet; gently elevate, and thereby promote gayety, and chase anxiety and care; warm the heart, and at the same time stimulate the flow of ideas, whence will come sociability, and with sociability, politeness and toleration, elegance and good taste.   It will prevent and cure dyspepsy, the most American and the least French of all diseases that scourge the world — in fine, by virtue of its tonic and stimulating properties, touch every weakness for which tonics and stimulants are prescribed —not, however, as a medicine, to lose its power with use, or be followed by reaction, but as a continuing condition — a habitual alimentation, like pure air, nourishing food, exercise, and proper clothing.

E 2

CHAPTER VIII.

EPERNAY.

I TURNED aside from my intended visit to Champagne on learning vintage was ended in that province, and went on to Paris. Two months later I ran over to Epernay, one of the chief seats of the commerce in sparkling wines, and presented myself to my old correspondent, M. Girbal, who received me• like a brother, and very soon put me in the way of seeing all worth seeing in the neighborhood. There are two cities at Epernay—one above ground, of buildings two and three stories high, and another under ground, of cellars two and three stories deep. This last, however, is not, like the Catacombs beneath Paris, a city of the dead, a receptacle of skulls and cross-bones, but a store-house of well-corked and wired bottles, full of pent-up life and sparkle, laughter and noise. The caves I visited, and which took the whole day to explore, were those of Moet and Chandon, Piper and Co., Ruinart, and Roussillon. The first of these I found the most extensive, and the

last the most interesting; for these M. Roussillon himself showed me through, and voluntarily gave such
full and frank explanations as stripped of nearly all
its mystery an art whose few professors in America
seem to keep it as close a secret as if it were alchemy.

Very little masonry is seen in the cellars of Champagne. Except an occasional patch of brick or stone
to fill up a fault in the natural formation, all was
hewn out of the solid chalk. Easily cut as this is,
it is nevertheless abundantly strong, and durable as
rock, while its chemical quality seems to render the
atmosphere of its chambers singularly pure and dry.
A two-story cellar is common, and some are even
three deep. Mad. Pommery, of Rheims, is making
one, I am told, of which the floor of the *upper* story
will be eighty feet below the surface of the ground.
Were it not for the ease with which Champagne bot-
tlers can burrow in the earth, their wine could not be
afforded so cheap as it is. To construct of stone or
brick caves as vast as those, for instance, of Moet and
Chandon, would require so great a fortune that upon
the interest of it both Moet and Chandon might live
like princes.

The wine grown in Champagne is a natural sparkler. With Catawba, Burgundy, Hock, and all other
sparkling wines known to commerce, the fermenta-

tion which ensues immediately on the first bottling having done its work in developing the gas and depositing a sediment subsides, and is never heard from again ; but with true Champagne new fermentations repeatedly occur, each one depositing its sediment, to be got rid of by a fresh tabling and shaking. For instance, M. Roussillon showed me a stack of fine Still Sillery bottled, not to sparkle, but to keep quiet, and therefore without any addition of sugar, yet it had fretted and fumed within the glass during six or seven years before it would be Still Sillery. Two and often three disgorgings and recorkings are needed before it is safe to send out for sale. By reason of this foamy quality it is that makers of Sparkling in other parts of France often use a certain portion of wine grown in Champagne to mix with that of their own districts. I am inclined to think, from my experiments with it, that the Scuppernong, produced in North Carolina, is as good a natural sparkler as we need.

Usually at least three qualities, growths of different places in the province, are mixed together, which is done toward the end of December following vintage ; but the finer kinds are never mixed. And my entertainer, M. Girbal, out of consideration for his health, puts up what he needs for table use wholly unmixed, although not using, he said, raw wine of very high quality. I can say for it, however, that it was good, and had a fresh and free taste, more like Sparkling Catawba than any Champagne I had drunk.

They make four kinds of Sparkling—high sparkling, common sparkling, half sparkling, or *crémant*, and tisane. The half sparkling is best, and the tisane the most inferior. But better than all, and the true type of Champagne, is that which does not sparkle at all, being entirely free of sugar or other admixture, and bottled when new merely in order that while ripening it may keep its fresh and delicate flavor. And this is the original of bottled Champagne. The plan of forcing a sparkling fermentation only gradually grew out of the ancient practice, which did not aim at producing foam and noise, but only at preserving purity, delicacy, and grape-blossom bouquet, that they might become united to maturity and fineness, like a wedding of youth, innocence, and beauty with experience and wisdom. Of such is Still Sillery, and I can testify that some M. Roussillon gave me was delicacy itself and purity itself.

For Sparkling wine an early vintage is considered important. The fruit is put to press soon as may be after being gathered, with no crushing whatever, and in as solid a condition as the necessary handling and transportation will permit. The juice reposes in large casks or vats from twelve to twenty-four hours, to deposit its coarse lees, after which it goes into new casks of moderate size. These they prepare first with a washing of hot water, and then, after drying them, with a fumigation of burning sulphur, or, what I prefer, of burning brandy, flung into the cask in the proportion of a gill to every barrel of capacity, and lighted with a wisp of paper. Late in December the mixing takes place, which is made the subject of much deep study and discussion—this sort being put in for sparkle, this for body, this for bouquet, this to prickle the tongue, and this for quantity. After the mixing comes the clarifying, performed by stirring in isinglass dissolved in older wine. Russian isinglass is the best. Then comes a medication with nutgall and alum in no small doses (*dose* is the French word for all the doctoring wine receives). Toward the end of March or early in April a second drawing-off takes place, accompanied with filtering through a sieve having two bottoms, one of hair and the other of silk. Soon afterward the bottling, which must be accomplished before the first of September, may begin.

A body of wine, to the quantity usually of many thousand gallons, is brought together in one or more large casks. About two thirds of the whole is new, and one third old. At this stage the decomposition of the sugar contained in the must ought to have exhausted three fourths of it. Of the natural sugar thus remaining, and what is afterward to be added in the form of rock-candy, the wine should contain, when it goes into bottle, the quantity of 7 pounds to every 225 bottles. To ascertain the true proportion to add, the following is an approved method:

Take fifteen pints of the wine, and slowly and carefully boil it down to two pints and a half. Twenty-four hours afterward test it with the *gluco-œnomètre*, as they call the wine-scale. If 5° below zero of the scale is indicated, it will not sparkle even at from 68° to 77° of Fahrenheit. In such case add, in the way hereafter described, 7 pounds of white rock-candy for every 225 bottles. Should the wine show 6° on the scale, then add, in the same way, 6 pounds of candy; if it shows 7°, add 5 pounds; if 8°, then 4 pounds; if 9°, 3 pounds; if 10°, only 2 pounds are needed; 1 pound for 11°, and for 12° nothing at all.

A simpler plan was devised by a peddler of wine-scales: Float the scale in a quart of the wine, and if it falls below zero, add sugar, carefully measured and mixed, until you bring it up to zero. This gives the proportion needed.

The sirup, called "liqueur," is composed as follows:

For a sixty-gallon cask of sirup, take 300 pounds of white rock-candy, as pure as can be had, and 2½ gallons of fine Cognac brandy, and fill up with wine more than a year old. Every day, for twenty days, roll the cask well, so as to dissolve the candy; then filter and bottle, to keep till needed for use. The sirup is mixed with the wine eight days before the latter is bottled.

Once got into bottles, and corked and twined, the wine is allowed to remain on the ground floor, where the temperature should be from 68° to 77° Fahrenheit until the fermentation has got headway enough to break the glass merrily, when it is removed to an arched cellar, where the temperature ought to be from 50° to 52° Fahrenheit, there to remain till next year, when it is brought up and stored in an intermediate cellar, whence again, before going into market, it must be removed to a store-room above ground, to become tempered to the exposures it is to undergo.

The bottle fermentation, in consuming the sugar, develops carbonic acid gas and alcohol, and deposits a sediment. To get rid of this last, the bottles are placed in racks, in which are holes to receive the neck and support the shoulder, and so formed as to allow them to take any position, from one nearly horizontal to one nearly perpendicular. Every day they are shaken with a twisting movement, designed to gently detach the crust of sediment without troubling the liquid; and at every shaking are changed in position, till from one nearly horizontal they are gradually brought to one nearly upright, bottom upward. By this time the sediment is entirely gone from the side, and rests against the cork. This operation requires from fifteen to twenty-one days, and can usually be performed at any time after February of the first year.

When it becomes necessary to prepare the wine for market, the operation of *dégorgement* takes place. Holding the bottle carefully, the workman, with an instrument half hook, half knife, cuts the lacing; the cork, sometimes coaxed a little with the thumb, flies out, followed by a gush of sediment and froth. Wine would flow but that the neck is raised in the nick of time. Then, tapping the butt lightly with his hook, he starts a further outpouring of froth, and, as it comes, rubs with a finger the inside of the neck, to help the foam wash away all adhering sediment— and the problem is solved, that might have puzzled a conjuror, of how to remove the sediment from underneath the wine without disturbing it.

Then, if the wine is deemed fit to market, comes the last dosing of sirup, intended to give the proper

taste. If there is nothing wrong about the wine,
nothing to be helped, nothing to be masked, the
sirup I have described is all that is needed for the
last, as well as first sweetening. But a good deal
more is usually added, and it is the composition of
this last dose which is *the* secret and mystery of
Champagne. Here is one recipe for wine intended
for the English market:

Port wine,	Cream of tartar,
Cognac spirit,	Sugar,
Cognac brandy,	Kirsch,
Brown Cognac,	Raspberry extract,
Elder-berry juice,	Madeira wine.

The sirup having been carefully dosed in even
quantity to every bottle, new and better corks are
driven in and wired down, the bottles are moved or
waved about in a way that mixes well the contents,
but can not be called shaking, and in a few weeks,
more or less, may be sent off.

I was more than ever convinced, by what I saw at
Epernay, that if we would make Sparkling wine in
America, we must first make the makers of it, and
not import them ready-made from abroad. What
chef de cave from Rheims or Epernay, for instance,
to whom you might give 1000 gallons of Catawba to
bottle, would not begin by preparing it with nut-gall,

tannin, and alum, to correct a disease called graisse, which I never yet knew the Catawba or any other of our wines to have, and which, in consequence of its excess of tartar, I am sure it is impossible for it to have.

If he were very sapient, he would also undertake to mingle different kinds, a thing quite unnecessary, and, as regards effect on the health, more or less pernicious, though in time we shall probably come to it. He would be pretty certain to add to the new wine a certain proportion of old, for he would not know that wine ripens here faster than in France, hence that no such mixture is either necessary or proper. (For my part, I consider it highly injurious, where Catawba is the wine, and would be slow to believe it good for any.)

Then he would have his secret recipe for the sweetening sirups, which he would as carefully conceal from his employer as if it were actually the philosopher's stone; but which, could he be induced to reveal it, would prove to be something like that I have just given, and which, especially designed for John Bull's palate, sweetens his posset with Port, Madeira, and Cognac, Cognac, Cognac, etc., whereas, in fact and in truth, so far as relates to bottling wine grown in the Ohio Valley, every drop of spirit added is a positive injury, and, except perhaps what little may be required to preserve the sirup until needed for use, has, after full trial, been abandoned.

By discarding these dosings and mixings, we may thus get rid of the most troublesome and complicated part of the business; very little seems to remain that we may not learn for ourselves. In the beginning we shall blunder, it is true; but a Champagne or Hochheim professor would blunder worse. We shall have to learn; they would have to unlearn as well.

One great obstacle in our way is the difficulty of obtaining good, reliable bottles and corks.

The reason why I have not gone more fully into details is that the Sparkling wine business is so hazardous, and the capital that must be hazarded is so large, I shrink from the responsibility of helping any one to embark in it. Besides, any who might undertake it would find it quite as easy to obtain from abroad all the best treatises on the subject, both German and French, as to buy and import reliable bottles and corks, and the latest and best machines. I have only been trying to clear away, for the benefit of beginners, some of the cobwebs of mystery woven in the cellars of Champagne, and which my visit there helped me to see were only cobwebs.

The Americans love pop, foam, and noise, and will

always consume largely of gaseous drinks. They have in the Catawba a wine capable of great things. Let but the product be large enough to allow the bottler to select only the choicest specimens, and of the best vintages, and those who follow the business properly, and especially those who secure good corks, need fear no competition from any thing *likely to be sent over here*, however it might be if the comparison were with those princely qualities found only on the great tables of Europe. There are those who think the day of the Catawba has gone by, but I am not one of them. Its wine has qualities which peculiarly fit it to combine with sugar, either in the bottle or the "cobbler." The last, made of new and sufficiently acid wine, such as is easily found in the West, but seldom or never in the East, is a summer drink of unsurpassed excellence. Certainly there is nothing in Europe to match it. Many an American traveler would be glad if there were, and be glad, too, if he could exchange the best grapes of foreign fruit-markets for the clusters he loved at home. In its place I will consider the question whether there is danger of this valuable variety being destroyed by the oïdium.

We will visit Champagne again when the leaves are green on the vines, and bestow our time, not on the dark, deep cellars, but on prettier objects above and outside of them.

CHAPTER IX.

PARIS AND THE GREAT EXHIBITION.

A GREAT Exhibition was that of Paris in 1867
—grand and magnificent as a battle—and a
battle indeed it was, wherein not two, nor three, but
all the nations of the round world met in the contest, and mingled in the display

" Their rival scarfs of rich embroidery."

Worthy was it to be remembered by all who bore
part in it, as a soldier remembers he was present and
fit for duty when some proud day of war was won.

In the three months and a half during which I
was an almost daily attendant in the Champ de
Mars, I saw a good deal of liquor consumed. Every
country had its restaurant, where the drinks native
to its soil were drunk by the natives of others—a
pleasanter way, that, of tasting the soils of distant
lands by sample, as it were, than of acquiring a
knowledge of them by dint of locomotion.

The American restaurant dispensed soda - water

and iced drinks — exclusively American I believe they all are—which not only astonished, but delighted multitudes, who took the first glass from curiosity, but the second from appreciation. An Englishman who had carried either his curiosity or appreciation as far as the tenth glass, said, "Your people ought to excel in compounding drinks, for, taken by themselves, your liquors are infamous." "Yes," I answered, "necessity was probably the mother of their invention. Nature having thus far denied to us those more natural drinks with which other people are blessed, we have been forced to imitate them with what materials were at hand. 'Cobblers,' 'juleps,' and 'cocktails,' 'stone fences,' 'hail-storms,' and 'smiles,' are but so many different kinds of American wines. There is spirits for strength, sugar for taste, lemon-peel or mint for bouquet, and powder of ice for quantity."

And why are they *not* wines? and why should we take the trouble to grow wines when the bar-keeper can so easily and deftly make them for us? Is tannin wanted? nut-gall is cheap; or, Is color desired? elder-berries and logwood are still cheaper. If the disciples of Chaptal are right, who say their imitations effected by fermenting quantities of sugared water on a few grape-skins and seeds, or mixed with a certain portion of true grape-juice, are really wines, then those who bring together in the tumbler spirits, sugar, water, and an aromatic, without going through any manœuvres at all of viticulture or vinification, are likewise wine-makers, and we, God bless us! are a nation of wine-drinkers.

But are these imitations really wines to all chemical intents and purposes? or, if they be, are they likewise so to *all* intents and purposes, and all effects and consequences as well? Chaptal, who was minister of the interior under Napoleon the First, and was, moreover, a great chemist, did not push his theory farther, I think, than to recommend additions of water and sugar to the must of imperfectly ripened fruit, and that to no greater extent than would make three barrels out of two. But Doctor Ludwig Gall, of Germany, whose recipes for falsification our government has taken pains to promulgate, through the Patent Office Report of 1860, goes farther, and obtains a double product. His theories seem to be well reasoned out, and his results have become so acceptable in Germany that, as he informs us, falsification has been for many years in general practice there. These theories, having again been made known and advocated in the lately published work of Mr. G. Hussman, of Missouri, who, as disciples are apt to do, goes farther than his master, the time has come for accepting or refuting them.

Gall finds the useful element of wine to be diluted alcohol (and in warm countries tannin also), and its attractive elements to be bouquet matter and flavoring matter derivable from grape acids; and he makes little or no account of any thing else.

In fully ripened grapes he finds from 28 to 30 per cent. of sugar, which is just enough to give the right proportion of alcohol — about $6\frac{1}{2}$ thousandths of acids, which is just enough to give the right proportion of flavor, and just enough bouquet matter to give the right proportion of aroma.

In unripe grapes he finds too much acid, too little sugar, and either too little bouquet matter, or none at all.

To make a good middling wine equal in all things except bouquet to any obtainable from fully ripe grapes, he dilutes the must with water till the acid is reduced to the true proportion of $6\frac{1}{2}$ thousandths, and then adds sugar until the whole quantity of sugar is increased to the true proportion of 28 or 30 per cent. Bouquet he does not attempt to supply.

Making the quantity of acids found in his must his base, and having by tests ascertained what that quantity is, he is no more at a loss for his other in-

F

gredients than Dr. Sangrado was, who cured every thing with warm water and the lancet: he brings both pump and sugar-hogshead into requisition, and makes sure of full a thousand pounds of wine for every $6\frac{1}{2}$ pounds of acid, which is often twice as much as his grapes could have made in the natural way, thus producing more wine in a bad year than in a good one. Such a vine-dresser needs little of the smiles of god Bacchus—who was the original Sol, I think—but should rather pray to him for clouds and rain, that quantity may be abundant and quality middling. And, could he find a grape that would never ripen at all, his fortune would be assured. I don't know why the grape acids might not be manufactured from some cheap substance, as the grape-sugar Gall uses is from potatoes; and then the vine and its fruit might be dispensed with altogether, and science triumph!

Gall goes further. Finding a good deal of acid substance remaining in the pomace after pressing, he obtains all of it he can by soaking in water, establishes the true proportion of $6\frac{1}{2}$ thousandths or less, fills in sugar to the quantity of 16 per cent., ferments, then adds of *spirits obtained from grape-sugar*, that was once potato starch, *eight per cent.*, and has a wine as good as the other, and even a better, if the grapes were ripe, for the pomace affords not merely acid, but bouquet as well.

In the larger part of the American vine districts grapes usually ripen too well to furnish the excuse, and, at the same time, the foundation that Gall and his friends in Germany enjoy, namely, excess of acids. So his disciple here, Mr. Hussman, finds a new foundation in the excess of bouquet matter with which most of our grapes are afflicted, as well in good as in bad seasons. Making bouquet matter, then, his base, and paying little attention to the acids, he estimates the quantity of dilution it will bear, and manfully pours in common water and cane-sugar till he runs his product up to a point beyond what even his teacher dared aim at. He, too, insists his wine is as good as the original; yes, and better too. Less discreet than his neighbors in Hermann and its vicinity, who, he thinks, will blame him " for letting the cat out of the bag," they preferring to devote themselves in secret to the pursuit of the new science, he glories in the discovery and its results. Let him be heard.

" But let us glance for a moment at the probable influence this discovery will have on American grape-culture. It can not be otherwise than in the highest degree beneficial; for when we simply look at grape-

culture as it was ten years ago, with the simple product of the Catawba as its basis, a variety which
would only yield an average of say 200 gallons to
the acre—often very inferior wine—and look at it
to-day, with such varieties as the Concord, yielding
an average of from 1000 to 1500 gallons to the acre,
which we can yet easily double by Gallizing, thus, in
reality, yielding an average of 2500 gallons to the
acre of uniformly good wine, can we be surprised if
every body talks and thinks of raising grapes? Truly
the time is not far distant—of which we hardly
dared to dream ten years ago, and which we then
thought we would never live to see—when every
American citizen can indulge in a daily glass of that
glorious gift of God to man, pure light wine, and
the American nation shall become a really temperate people."

"And there is room for all. Let every one further the cause of grape-culture. The laborer, by
producing the grapes and wine ; the mechanic, by in-
ventions ; the lawgiver, by making laws furthering its
culture and the consumption of it ; and *all*, by drinking wine, in wise moderation of course."—Page 172
of "GRAPES AND WINE."

Truly our German friend has large notions of what
he terms drinking in moderation. Let us see how

large that daily glass of the American citizen must be to hold all that is about to be poured into it.

On page 22 of the book we learn that at the time it was written there had already been planted 2,000,000 of acres of vineyard, or two fifths of the area devoted to vines in France. These, when in full bearing, as they will be in 1870, should, according to his lowest estimate, yield 2,000,000,000 gallons of natural wine, about twice the product of France, from which, by Gallizing, we shall obtain 5,000,000,000 gallons. Excluding children too young to drink, there should be in the whole country some 25,000,000 able-bodied drinkers, the share of each of whom in the yearly vintage would be 200 gallons, something over three bottles a day. Then there are the teetotalers—they might object to drink their share; but I suppose we might funnel the teetotalers.

When Mr. Hussman wrote, American wines were selling at wholesale for $2 50 per gallon; but since then, from increasing production, they have fallen to about $1 25, though consumers have to pay at least $2. The yearly crop, therefore, which we are led to hope for, will cost the drinkers of it $10,000,000,000, or sixteen times as much as we have of green money, plentiful though it be as forest leaves. With such a volume of wine to " carry," however, it is no wonder

there are people who think we still have too little of
that foliage of the root of all evil.

Of a truth we are ruined!

But both Gall and Hussman must give way to an
enterprising Frenchman, M. Petiot, who, for the last
ten years, has been working the same rich vein.

Gall makes as much wine as his acids will flavor.
Hussman makes as much as his bouquet matter will
odorize.   Petiot, taking a collective view of things,
assumes the must to contain 99 per cent. of water
and sugar, and only one per cent. of all other substances—tartar, tannin, resin, coloring matter, essential oils, and all.   "It is this one hundredth part,"
says he, "to tell the truth, *which constitutes the wine*,
which distinguishes it from other liquids, and which
principally gives the distinctive qualities and fixes
the price."

Having brought his wine matter into this small
compass of one per cent., he strikes out boldly, and
does not stop till he has attained a *five*-fold result,
having made, as he assures us, out of a quantity of
grapes, sufficient to give only fifteen hundred gallons
of natural wine, seven thousand gallons of the chemical article; all of it, he asseverates, better than the
original, and with a better bouquet.

One thing is strange.   Chaptal flourished sixty

years ago. Gall and Petiot have been illuminating their respective countrymen some twelve years or more, and yet the plantation of the vine is every where extending, the natural product augmenting, and, at the same time, the price yearly rising. I knew great truths made their way slowly, but did not know great falsifications did.

Graft Petiot on Hussman, and our crop in 1870 will be something like 12,000,000,000 gallons, obliging each of us to swallow eight bottles a day, and to pay for it the very pretty figure of $24,000,000,000; and where is *that* money to come from, one would like to know?

And enthusiastic Hussman calls on us to persevere in the good work, and extend the culture more and more! Surely

"A Dutchman's drink must be
Deep as the roaring Zuyder Zee."

Gall has offered a premium to any chemist who will detect in his brewage any thing hurtful to health, and cites high chemical authority to the effect "that no substance conducive to health is *removed* from the wine by an addition of sugar and water before the commencement of fermentation." The others are equally certain there is nothing unhealthy contained that chemistry can detect.

This is precisely the claim of those who feed cows on distillery slops concerning the milk they sell. And chemists find nothing in it but the water, sugar, butter, caseine, and salts proper to the milk of all cows, only they are combined in somewhat different proportions; and yet the children die of it, many and fast. There is a limit to the authority of chemistry in regard to aliments. Who would like, for instance, to eat a chemically-compounded egg, having every quality of the hen-laid article which analysis could detect? But is it true these simulated wines are chemically identical with the real thing?

The sum of the theory would seem to be that wine is diluted alcohol agreeably flavored to the taste, and sometimes perfumed also—a fair definition of a mint julep.

Here are, 1st, alcohol; 2d, acids; 3d, bouquet matter; and, 4th, water. All these exist in real wine, and all exist in the false as well. Admitting, for the moment, these to be all the ingredients contained in either, let us look at them separately.

1. Alcohol. Not only do they produce this element by fermenting sugared water, but where none of the must is used, and all the wine matter is extracted from the pomace, Gall enjoins adding distilled spirits in the proportion of eight per cent.

Concerning such adulteration with spirits, Mr. E. Delarue says : " Take some of the wine to be tested in a porcelain capsule, which place over an alcohol lamp. Float in the wine a nut-shell filled with oil, in which put a floating taper. Ballast the nut-shell with shot till its edges are brought even with the surface of the surrounding liquid. Light the lamp and the taper. Now, if you place a thermometer in the bowl, you will see that at 45 degrees of *Centigrade*, alcoholic vapors will rise from the wine and catch fire, forming round the taper a reddish halo. Repeat the experiment with natural wine, and the vapors will not show themselves until the wine has reached 90° of Centigrade, almost boiling point. In the first place, the alcohol was in the condition of a simple mixture; in the second, it was in a state of combination, or, we may say, *intimate incorporation*, and retained by a cohesive force not to be broken except by a high degree of heat."

The true thing, then, adheres till the heat reaches 90° of *Centigrade*, almost boiling point, while the imitated thing lets go at 45°, going off at just " half cock." Now the normal temperature of the stomach is from 98° to 100° *Fahrenheit*, only 13 to 15 degrees below the point at which distilled alcohol separates itself from other ingredients with which it

may be mixed. Must it not, then, when taken into
the stomach, almost immediately explode upon the
nerves of that organ the whole of its stimulating
power, to be as rapidly communicated to the brain?
while undistilled alcohol, on the other hand, as it exists in true wine, bound to its associates by the hand
of Nature, and not stirred in with a stick, requires
twice as much compulsion to make it part company,
works but slowly while in the stomach, passes out of
it in a combined, qualified, and modified form, and
so, entering into circulation, expends its force gently and slowly (may we not presume) upon each and
every part in such way as that all are equally ex-
alted, thus preserving equilibrium instead of disturbing it, as every agent must which exhausts itself upon
only one or two organs.

It is a fact that a person whose stomach is sensi-
tive can easily detect the presence of distilled alcohol in wine by the burning which he feels after
drinking it.

This may be the proper connection for suggesting
that the slower effects of red wines as compared with
those of white may be due to the enveloping, so to
speak, of a considerable portion of the alcohol by the
coloring matter, which, being a resin, will dissolve in
alcohol, but not in water.

Liebig says: "Owing to its volatility, and the ease with which its vapor permeates animal membranes and tissues, alcohol can spread throughout the body in all directions."

Evidently the quickness or slowness with which so volatile a liquid passes to the state of an all-permeating vapor, to flash like thought from part to part, are most important when we are judging of the qualities of alcoholic drinks.

Spirits and water, whether in form called cocktail, julep, or punch, are, in this respect, just like Dr. Gall's brandied wines. If, by reason of the earlier decomposition of their alcohol and its too sudden and necessarily unbalanced action on the organs, juleps, cocktails, and punches are hurtful to health, happiness, and morals—if their tendency is to breed in the nervous system a disease called the drunkard's thirst, which true wine rarely does, then Gall's wines, holding eight per cent. of added alcohol, are, for the same cause and in the same measure, injurious to health, happiness, and morals, and equally productive of the drunkard's thirst. And, since nothing in a compound can be called a good ingredient unless it combines properly, an ingredient that goes loose at precisely the time when it should not, must be esteemed a bad one.

Chemists like Gall and his supporters, if he has any among chemists, who take no account of the manner in which the ingredients of wine combine with each other, as influencing their effects on health, are of small value as judges of another question now arising, which is, Do wines made of a certain quantity of grape-juice, mixed with a large body of sugared water, but with no addition of distilled alcohol, contain ingredients hurtful to health? Of this we are certainly at liberty to judge for ourselves, though I am as yet unable to indicate any chemical test bearing directly upon it. A presumption against alcohol developed by fermenting sugar and water arises from this, that, as in the case of distilled alcohol, which we have just been considering, the intruding ingredient comes in form of something artificially separated from matters with which it was once naturally combined. If the wine with which Mr. Delarue experimented held its brandy but feebly in combination for the reason that it had been separated by distillation from other wine to which it was native—and we can imagine no other reason—then, by analogy, we may infer that spirits developed by fermenting in water grape-sugar, extracted from starch that was itself extracted from potatoes, will be held as feebly by the water, etc., in whose company it happens to find itself, as the distilled alcohol of Delarue's experiment was by the wine into which it had been stirred. Extracts and mixtures naturally provoke suspicion. The sugar which ferments in juice of the ripe grape was always there. It and the watery particles of that juice can hardly be called sugar and water. They are one—born of one root, and kindred of one sap. Sap is thicker than water.

We have seen that Gall finds the true mother of wine to be its acids; that Hussman thinks bouquet matter is the real quintessence, though without expressly discarding the acids ; and that Petiot puts all virtue and wine power within the compass of the $\frac{1}{100}$th part, and dilutes it " *à discretion.*"

Do these materials, one, any, or all of them, properly combine with pump-water? Especially do they combine as intimately in the imitation wines as they do in natural ones?

Tartaric acid, in which German as well as most American wines abound, is said never to be present as a free acid in French wines. But it is used in France to adulterate with. To detect its presence, Mr. Delarue gives us the following test:

"Mix some of the wine with twice its volume of chloride of potash, saturated at the temperature of 15° *Centigrade.* Stir well with a glass rod, and if

in seven or eight minutes a crystalline powder of bi-
tartrate of potash is precipitated, the tartaric acid has
been added.   If it were natural to the wine, it would
not precipitate itself *under several hours of stirring.*"

Here we see the want of cohesion even more strik-
ingly indicated than in the case of the intrusive alcohol.   In place of stirring in some pounds of crystal-
lized tartaric acid, Gall and his believers collect a
mass of grape-skins and seeds, and stir them in, or
they bring together grape-juice in which the acids
already exist, and sweetened pump-water, set the two
to ferment in company, and ask the wine to share
half its quality with the water.   And they mix, it is
true, but how?   Why, in the same loose way as the
tartaric acid of Delarue's experiment did, to separate
ten or twenty times earlier and easier than acids naturally present would.

As regards the effect on health, would this be ten
or twenty times too soon?   Yes, if the substance
mixed with the sugared water by fermenting it on
grape-husks or with grape-juice play any part beyond
merely pleasing the palate.   They may be, and probably are, designed to qualify the action of the alcohol, of the water, and of each other, while passing
through the channels of circulation.   In such case, we
may reasonably suppose, it is as important for them

to combine well as to combine at all. If they have uses to perform together, they should remain together until those uses are performed. If they have uses to perform separately, they should separate soon enough to perform them. A drink composed for the use of man, and destined to circulate and decompose within his body, should not merely contain good ingredients, but they should hold together just long enough, and separate at just the right moment.

These considerations go to all those substances found in Petiot's $\frac{1}{100}$th part of pure wine, and which he says " constitutes the wine," including of course Gall's and Hussman's acids and bouquet matter.

Judging by Mr. Delarue's test, they are as easily shaken from the rest of the fluid they are designed to disguise as the lion's skin was from the shoulders of the ass, and for the same reason—they did not *grow* there.

And the water: has chemistry found nothing in the pump-water wines which can injure health, or which pure wine should not contain ? It has found there precisely what was in the well from which the water was pumped—lime, magnesia, clay, and whatever other impurities the earth through which it came could supply. Whatever impurities of this kind find their way into grape-juice are in a short time almost

entirely flung down.. But their quantity is never
very considerable in pure wine, as appears from the
comparatively small precipitate deposited when it is
brought in contact with oxalate of ammonia.

It is true, distilled water may be used, but it is
not, and will not be—that's all. Nor will very much
sugared water be fermented to produce the alcohol
required, though sugar for sweetening will no doubt
be put in. Whisky, or neutralized whisky, will be
found cheaper and readier.

It seems plain that these gentlemen have no other
idea of wine than as a diluted alcohol flavored with
grape acid, and sometimes, too, colored with grape-
skins. Will such a drink wean us from whisky and
rum, and make us a sober people, as wine-drinking
peoples mostly are? Neither the one nor the other,
I think. It is repugnant to common sense that men
can learn to love a mere chemical product as they
can a natural one. Cocktails, cobblers, juleps, and
punches are sweeter in the mouth than imitation
Catawba or Concord can possibly be made. And if
the copious supply of water with which we adulterate our whisky and rum has thus far failed to antidote the drunkard's thirst, the whole volume of the
Missouri or the Rhine, mingled with spirits from
cane or potato sugar, in however nicely adjusted proportions, will equally fail.

We are now yearly consuming, and being consumed by, 80 million gallons of whisky, mixed with a reasonable proportion of water; and, though millions of our people take the mixture regularly and copiously, it is precisely they who are either actually drunkards, or going to be drunkards, or in great danger of being drunkards. Even the teetotalers seem in a better way to keep sober.

Fortunately, sugar costs something. Hussman estimates the expense of making Gallized wine at 60 cents a gallon; a good deal more, strange to say, than he does the outlay for growing real wine the natural way. In the south of France Gallization is never dreamed of, for they grow their alcohol cheaper on the vine than they could brew it from sugar of any kind. And, when our countrymen shall learn to produce good, pure, wholesome wines at a cost of ten cents the gallon, there will be no need for my writing homilies at this prosy length upon the virtues of purity and truth.

AMERICAN WINES ABROAD.

Our Sparkling wines found greater favor on the palates of the jury which tried them than did the others with that which judged of them, which has helped me to believe in the excellence of the Catawba in the sparkling form. It was rather too much to ask of the French members especially, to fall in love with what seemed to them new and "*sauvage*" aromas and flavors. The German members appreciated them higher, and more justly, I am sure. The Norton Seedling, red and not pink, fully fermented on the skin, was received with decided respect, and complimented on its comparative *neatness* on the tongue and palate.

We presented to the juries some 90 different samples. Most of the Sparkling received honorable mention, the others no mention at all. I hope we shall make a better show next time.

The magnificent displays of the French, Bavarian, Austrian, Portuguese, and other departments afforded me an opportunity to make acquaintance with the products of the great world of vines in all its districts. The loss of a memorandum-book, however, prevents my giving some statistics in this connection that would be interesting.

PORT WINE.

I had read in the *London Times* a communication, to which a conspicuous place was given, which argued that Port wine, as produced in Portugal for the British market, was brandied up to the very high point of 26 per cent. of pure alcohol, for the sole purpose of making it keep. While tasting Portuguese wines, I asked the commissioner if this was true. He answered no. The addition of brandy was to make it palatable and strong. To make Port as the British people love it, the fruit is first flung into an enormous tub or shallow vat, where as many as ten, twenty, thirty, and even sometimes sixty men, with trowsers rolled up above the knees, trample on it during from twelve to forty-eight hours, and until it is reduced to a thick mush. After this it is set to ferment, but at a certain point fermentation is arrested by adding brandy, which is done to preserve a portion of the richness of the grapes. Afterward the tendency to ferment is kept in check by repeated additions of spirit, until, from containing a moderate amount of alcohol, it becomes about as strong as brandy and water—half and half. The fourteen years required to ripen this mixture is needed, not for the wine, but for the spirit, which, it seems, wants more age in mixture than when pure.

VIENNA BEER AND PARIS WATER.

How the French took to the Vienna beer on draught in the Austrian restaurant! Well they might. No water in reach of Paris could make such as that. Paris beer is not fit to drink. The Europeans have the best excuse for abstaining from water. You rarely find a drop that you can take without present disgust and future trouble. From this cause it is that water is drunk usually in the form called soda-water, of Seltzer-water, which is, in fact, only the evil fluid of the Thames or Seine, crammed with carbonic acid gas. By far the best drink for an unfortunate teetotaler I found to be Eau de St. Galmier, brought from some distance, and costing four cents a bottle without the bottle. It has no impurities, except a good quantity of naturally incorporated carbonic acid gas, which, when sugar is added, makes it quite agreeable. The modes of supplying the cities are as defective as the quality supplied. The Seine water is filtered and sold by the bucket—which means that some of it is filtered, and a good deal sold by the bucket that is not filtered, but merely drawn from the hydrant in the courtyard. You pay for the pure article, but get the Gallized one, the concierge and your own domestics conspiring often to cheat you not merely in quality, but in quantity also. Now Seine water is notoriously impure. To attempt filtering away its impurity, so as to present it as a drinkable fluid, is a cruel joke. In London it is no better than in Paris. The waterworks of our American cities are a national glory, and their outflow, bright and pure, goes far to keep us in the sad habits of the Rechabites, Mohammedans, and Teetotalers. In the day when red wine shall come, their crystal wealth will still be useful at table to dilute table-wine for such as like it diluted.

The first of July drew near, and I was still in Paris. The long, slow, pea-green spring had passed into a beautiful summer of fair weather, warm days, and rather dusty breezes. I knew well enough the vine-blossoms had come and gone, and the fruit was swelling in the clusters, and that I ought to be among them. But the Great Exhibition was in Paris, and to it and Paris the four quarters of the world were being gathered in. One, finding himself there in those days, found it easy to remain.

The great day of all the seven months was the first day of July, when seventeen thousand of all races and nations assembled to witness the distribution of the rewards of merit at the Palace of Industry. On entering, I looked about to see where the fresh air was to come from, and found only a few port-hole openings beneath the vast sky-light, enough perhaps to ventilate a beer-hall, and they only half opened. The weather was hot, and yet the people lived. The Continentals know little of ventilation.

The only other fact left me to chronicle by the all-reporting press concerning the doings of the day is that, when the imperial party, in their tour of the hall, came opposite the American benches, there was a hurrah many times louder than had greeted them from those of any other nation, though the Americans were but few, which proves how much they love the great man, or else how much they love to hear themselves shout.

I admit I occupied a good deal of the time in looking with a field-glass at the imperial, and royal, and princely people on the stage, almost opposite to where I had the good fortune to sit, and especially in noticing the empress, as she bestowed on each one who mounted the steps to receive his medal at the emperor's hands an imperial bow and a Eugénean smile. Wonderful smile! as sunny-bright and honey-sweet as a landscape of Claude Lorrain, and, like it, a work of art too. She had just heard of poor Maximilian's death, but must for the day keep it a secret; and heavy as the bullet in his buried heart was the hidden burden she carried in her own, all the while she smiled and smiled.

PASTEUR AND HIS DISCOVERY.

I heard an unusual clapping of hands. Looking for the object of the applause, I saw mount the steps a gentleman whose embroidered coat showed him to be a member of the Institute. It was M. Pasteur, taking his great gold medal for having found a way to cure wine of all its diseases, and make it keep indefinitely long. With Pasteur was H. Marès, of Montpellier, who, better than any one else, I think, has taught us how to cure the oïdium and keep it cured. I have, since then, been so happy as to make M. Marès' acquaintance, and, when I come to recount my visit to Montpellier, I will introduce him to the reader.

M. Pasteur has published a large volume on the subject of his discovery. The important part of the work for the practical vintner to know is the instructions how to conduct the process of heating the wine, in which the remedy consists, and these I have tried to condense in what follows.

After explaining at length the nature of the various wine diseases, acidity, bitterness, etc., tracing them to vegetable parasites, and detaling his experiments in search of an agent to destroy the parasites, M. Pasteur arrives at the conclusion that they are effectively destroyed by heating the liquid up to a point between fifty and sixty-five degrees of *Centigrade*, which would be from 122° to 149° of Fahrenheit. This can be done in a "*bain marie*," that is, by placing the bottle or cask in a vessel of water, and heating the water; or by hot-air closets, or by steam-pipes introduced into the casks. The heating should be *carefully and gradually done.*

The following is in the words of the book: "The bottle being corked, either with the needle or otherwise, by machine or by hand, and the corks tied on like those of Champagne bottles, they are placed in a vessel of water; to handle them easily, they are set in an iron bottle-basket. The water should rise as high as the ring about the neck of the bottle. I have never yet completely submerged them, but do not think there would be any inconvenience in doing so, provided there be no partial cooling during the heating up, which might cause the admission of a little water into the bottle. One of the bottles is filled with water, into the lower part of which the bowl of a thermometer is plunged. When this marks the degree of heat desired, 149° of Fahrenheit, for instance, the basket is withdrawn. It will not do to put in another immediately, as the too warm water might break the bottles. To reduce the temperature to a safe point, a portion of the heated water is taken out and replaced with cold, or, better still, the bottles of the second basket may be prepared by warming, so as to be put in as soon as the first comes out. The expansion of the wine during the heating process tends to force out the cork, but the twine or wire holds it in, and the wine finds a vent between the neck and the cork. During the cooling of the bottles, the volume of the wine having diminished, the corks should be hammered in farther. The tying is taken off, and the wine is put in the cellar, or on the ground floor, or in the second story, in the shade, or in the sun. There is no fear that any of these different modes of keeping it will render it diseased; they will have no influence except on its mode of maturing, on its color, etc. It will always be useful to keep a few bottles of the same kind without heating it, so as to compare it, at long intervals, with what has been heated. The bottle may be kept in an upright position; no mould will form, but perhaps the wine may lose a little of its fineness under such conditions, if the cork gets dry and air is allowed too freely to enter."

M. Pasteur affirms that he has exposed casks of wine thus treated to the open air, on a terrace with

G

a northern exposure, from April to December, without any injury resulting.

Wine in cask may be heated by introducing a tin pipe through the bung-hole, which shall descend in coils nearly to the bottom, and return in a straight line, and through this pipe passing steam. If, after being thus once heated, there is such exposure to air, by racking or bottling, for instance, as to admit a fresh introduction of parasites, the disease may again be cured in the same way as at first.

M. Pasteur's theory, or practice rather, is well received in France; so well, in fact, that, though it was first made known in 1865, already, at the time of which I write, others had begun seriously to contest his claims to original discovery. The Vicount de la Vergnette-Lamotte, of Burgundy, insists that it is no other than his own method of storing wine during July and August in a garret, where the temperature might mount to 90° Fahrenheit, but perhaps not higher, while Pasteur heats up to 122° at least. It is said, too, that an unscientific person who lived in the last generation used to heat his wine to preserve it; and in the south of France they have a custom of exposing wine to the rays of the sun on the roof. A valuable invention is always attacked.

M. Pasteur's book has a very full report in favor

of his system, made by a committee of the Wine Merchants' Association of Paris. But all certificates are
dispensed with by such a scene as I have described,
where the value of the discovery is certified, and the
discoverer rewarded by a great round first prize gold
medal from Napoleon's hand, and a smile from Eu-
génie's lips.

CHAPTER X.

RHEIMS.

THE fifteenth day of July, 1867, found me lodged at an agreeable little hotel in the old city of Rheims, capital of the ancient province of Champagne, and next door to the great cathedral of coronation fame. I like such kitchens as that of the little hotel, close to the main hall and opening on it, where the cook is always present and ready to receive your orders direct, dressed in fresh white linen, with a white cap on his head, and surrounded by dozens of brilliant copper saucepans, while on the milk-white dresser is piled the provision for the day —chops, steaks, and joints. Such a sight prepares one to be pleased with his dinner in advance. He was as dignified—the white cook in question—as a chemist in his laboratory, and a great deal cleaner.

My old friend F., for ten years our *chef de cave* in Cincinnati, had returned some two years before with a snug fortune acquired in America, and was living at his ease in his native Rheims, or trying to do so. I had looked him up the evening of my arrival, and the next morning he called early, and arranged to take me a tour among the estates best worth seeing in the neighborhood. Soon after breakfast we drove out. As we left the town, the broad valley of the limpid and sage-green Marne lay before us, the foreground of the vine-covered hills, or rather mountains, rising with gentle slopes beyond.

"Is the soil of the plain rich?" I asked.

"No; it will bear tolerable crops, but needs enormous manuring. It is of chalk, like the hills; all about here is chalk."

A chalk soil is always pure, owing to its slow decomposition, I suppose. It is also fine. Champagne wine is born on chalk hills, and grows old in chalk cellars. The same formation as that I found at Rheims I had last seen in the form of the white cliffs of Albion, and before then in the South Downs, where the finest of mutton sheep graze on its short grass, and take their quality from its fineness and sweetness. The old rule was in force in the Marne Valley, as elsewhere—poor soil, rich product; great wine in little quantity.

Crossing the river and the level ground of its valley, the road conducted us with an easy rise up the

base of the mountain of Rheims, and soon we found
ourselves among fields of the most stunted and close-
ly-set vines ever seen, cultivated with unwearied pains,
and enjoying the choicest soil and best exposure —
vines of quality they. Higglety-pigglety the vines
of quality stood, each one with its little stake to help
it hold up its small offspring of fruit. They were
crowded much closer than those on the Côte d'or,
some plantations being set so thickly as to hold
25,000 to the acre. The object of this close planting in the one district, as well as the other, being to
force the ripening, by dint of suffocation and starva-
tion, at the expense of vigor.

A long drive among such fields brought us, at
length, to the quiet little village of Saint Thierry.
It was as still, in fact, as a deserted village. The
men were at work, the children at school, and the
girls and women were gathered about an old stone
fountain of abundant delivery, where, under wide-
spreading trees, they washed their linen sociably.
Under the trees we alighted, and went and looked
up the curé of Saint Thierry, to ask his services in
gaining us admission within the walled inclosure of
a great estate near by. The good man, after offering us wine, conducted us through the close-walled
and silent lanes, till we reached a great oaken gate,

which he soon got somebody to open. Entering, we found the grounds to be of most respectable aspect, tolerably kept, but pervaded by a lonesome, solemn, monkish air, easily accounted for when they told me it had been a convent once. Here was another fortunate coincidence. Did the good sisters choose this spot whereon to build their house of earthly tribulation because the earth about it grew so excellent a wine to moisten their clay, or was the good wine-soil brought to their doors because of their many virtues, and to reward their many abstinences? I have read that some of the nuns of Charlemagne's time had to be restrained from not very pretty habits, into which, in that rude age, they had fallen. He was a rough old fellow, and bad; but, like other old bad fellows, had a fine sense of maidenly propriety. He said he would be —— if his nuns should not be made to behave like decent women, and no longer stroll through the towns, haunt taverns, get drunk, and lie about in wayside ditches. How he shut them up, and how he punished them, I don't know. I can't think, however, that the vestals of Saint Thierry ever needed the emperor's rough discipline for over-draughts of the pure and sparkling wine grown within their own domain. It may sometimes have elevated their souls a little too near the skies, but could never have brought their bodies to the gutter.

From the walk to the chateau, I was attracted toward a balustered terrace on the left, flanked by two handsome flights of stone steps, from which there was a charming view. While enjoying it, a gentleman of good mien approached, and was introduced as M. Camus, the proprietor of the chateau of Saint Thierry, a resident proprietor, and not a Paris-haunting absentee, like too many in France.

Under his guidance, I took a long stroll through the vines, which was rendered very instructive by his conversation as we went. His vines were planted close enough to put 16,000 within an acre. They were all layered every year, and in the following manner: A trench is dug, usually nine inches deep at the foot of the plant, and extending from it to a point which the cane chosen for the purpose will reach after being pruned. Usually it branches toward the end, and the two branches, each cut back to two eyes, show above the soil, after they are laid down and covered up, like two slender and short twigs. Vines thus trained are called *low*.

There is also a custom of layering only once in seven or eight years, but then at such period all are layered. The practice of doing it only as plants grow sickly, or dead ones need replacing, which obtains in Burgundy, is not known in Champagne, I think. Vines laid down every seven or eight years are called *high*.

These need, of course, a trench to lie in suited to their larger size, and considerably deeper than nine inches—from sixteen to twenty, perhaps. Such vines are not set so thickly as low vines are.

I asked M. Camus if such frequent renewals, keeping the plants in perpetual youth, as it were, with no ancient and wrinkled stock to break the impetuous flow of sap, did not hurt the quality of the product. And I told him, in the same connection, of the old vines of the Côte d'or, and that in Médoc the juice of young ones was not allowed to class as quality No. 1.

He told me in reply that when new plants were set out upon the finer hill soils, which came from lower and inferior grounds, it needed some twelve years for them to acquire "the quality of the hill;" but as to vines that had acquired this quality, he did not think their yearly renewal affected injuriously their fruit. High vines were renewed only at each seventh or eighth year, and yet it was usually the low vines that produced the best wine. He thought old vines, though ever so often layered, remained old vines still, and were not rejuvenated to any purpose that could affect the quality of their fruit. Possibly he was right; but what becomes, then, of the facts that look directly the other way? Such contradictions are puzzling. Maybe old vines bear sweet fruit, not because they are wrinkled, or crooked, or ugly, but by virtue of some other quality, as yet unknown, that comes with age.

"You have but little oïdium in Champagne?" I observed.

"We have enough of it, though, and it has to be treated with remedies. My vines are never troubled with it, however. They would be if not frequently layered, I am sure." He then gave me details and reasons which convinced me he was right in this inference.

Like nearly all the proprietors of the district, M. Camus did no bottling, but sold his crop soon after it was made. He informed me that for twenty years, or thereabout, he had disposed of every crop to one house, always leaving it to the purchaser to fix a price. I regret that in losing my memorandum-book I lost some details obtained from M. C. of the cost of production, the average product, and prices obtained.

A view of uncommon extent and beauty was always before the eye at the chateau. The house, through portions of which they were good enough to show us, still preserved traces of the old convent. It was quaint, and yet rather grand too, inspiring a respect for its inmates. It would be as hard to reproduce as an avenue of old oaks. "All the modern conveniences" are convenient enough, I know, but they certainly don't impress the casual visitor with such a desire to stay and be at home as does the peculiar, stately, unpurchasable aspect of a house like that of St. Thierry.

The dwellers in such a home are to be envied, especially by a Western American vine-dresser, residing as they do in the midst of their possessions and effects—their wealth of vines about them and of wine beneath them—with enough to do and plenty to do it with—enjoying a farmer's freedom without his rusticity, and an aristocrat's luxury and comfort without his ennui. In America, to be a gentleman farmer requires a fortune at interest. Maybe the time will come when vine-dressing shall be so well understood and reliable for income—which it is not yet by a good deal—as that the owner of a hundred acres in grapes can sit under his own vines and trees in lordly independence and republican ease. As yet, however, chateau life is with us quite too near an impossibility, except in a suburban and imitation way. For the present, our fate is to be either rough-and-tumble farmers, or cits.

The next drive took us among high vines, which appeared, as to culture and training, much like those I saw in Burgundy. Both white and red grapes are grown, but, with slight exception, all are converted into white wine, and that is bottled for Sparkling. The white grape wine has more of the true Champagne quality, delicacy, lightness, and sparkle than what is obtained from red grapes, and is usually mixed with the latter in the proportion of one eighth, or one fourth.

The chalk or chalky rock of the vineyards is covered with but a thin layer of soil, which consists of about four parts of carbonate of lime and one part of silica and clay, which proportions are sometimes varied by the presence of oxide of iron.

At first planting the roots are not set very close, the distances being from three feet three inches between the lines by two feet ten inches within the lines, to three feet by one foot eight inches. Before planting, the soil is broken up to the depth of twenty inches or a little more. Two and three year old plants are preferred. Composted manure is flung in about the roots on setting out. The first year they are four times weeded; the second, they are cut back to one or two eyes, are hoed in March, and afterward weeded three times.

In April or May of the following year, a sufficient number of the most vigorous shoots are laid down to fill one third of the unoccupied space between the vines as they stand in their rows. The next year another third of the space is in the same way filled, and another year's layering completes the plantation with an utter rout (*déroute*) of the original ranks. At each of these layerings compost is freely used.

Pruning is done in February or March. From the middle of May to the middle of June they again stir the ground to the depth of about three inches. After blossoming, which usually comes about the 24th of June, the shoots are tied up, and buds not wanted are rubbed off. After this comes pinching in and another hoeing. Often there is a last pinching in as late as September, followed by a superficial hoeing, so managed as to remove from beneath any bunches touching the ground enough soil to let the fruit hang free.

Vintage often begins as early as the 15th of September, but the first week of October is oftener the time. The bunches are carefully examined by the "cutters," and all bad berries removed.

These details, which to some may seem inapplicable in our country, are given for the benefit of the colder portions of it, where grapes of desirable varieties do not ripen well, as at present cultivated. In the Marne, which lies on the northern verge of the European vine zone, fine and valuable wines could be produced in no other way. To the peculiar method just described the world owes the joy of that wondrous and sensational drink called Champagne.

If it is objected that American vines are of too rank growth to bear such dwarfing, let it be considered that the treatment is only for such as grow on meagre soils; that with more vigorous growers somewhat wider planting might do, until, by dint of continued dwarfing, they become smaller. I confess I would like to see the Delaware—choice little queen of pigmies—close set as above on the right soil and exposure.

The cost of cultivation need not be so great as many would suppose.

In some parts of the Marne a black earth containing sulphur is hauled upon the vineyards to improve the soil, but I did not happen to see it. They think it prevents oïdium. If there be sulphur enough, it is a good preventive; still, I would rather believe in layering, which, beyond doubt, is, when often enough repeated, a remedy of itself.

This reminds me that the remarkable success of grape-growing along the borders of Lake Erie is now

attributed by some to the presence in the soil of the
bituminous shale of the Hamilton group, such as is
found in the Pennsylvania oil regions. This shale
is said to contain, besides iron and sulphur, a large
quantity of potash. It is recommended by some to
be hauled upon sandy lands, as a manure to vines
planted there, and which are found to be much less
favorable to the grape than clayey lands, which, in
the region in question, abound in the shale.

The same Hamilton shale crops out in vast beds
on both sides of the Ohio River, from a little way
below Portsmouth down to about the mouth of
Brush Creek, in Adams County. I shall this year
try the experiment of spreading a layer of it over
a portion of my vineyard.

In many of the vineyards about Rheims sulphur
is in regular use. If layering be a perfect preventive on all soils, then the vines which need sulphur
must be high vines, laid down only every seven or
eight years. I am sorry to say I did not think to
ascertain this point.

In the garden of the house where Mr. F. lived
I was shown a large vine trained to cover a high
wall. One half of it was in good condition, laden
with fruit, and covered with dark green, healthy
leaves. The other half, on the contrary, had but lit-

tle fruit, and its foliage, faded to a yellowish hue,
was already falling.   Said F., "I tried, last year, an
experiment on that vine, by sulphuring well only
the half you see so healthy now, letting the disease
have its way with the other.   I did this to learn if
my neighbors in America were right, who say oïdium
can not be cured."

Evidently the result showed, first, that the oïdium
can rage in the Department of the Marne as well as
elsewhere ; and, secondly, that it is perfectly curable.
They apply sulphur three times in the course of the
season, and with a bellows and dredging-box.   All
with whom I spoke thought it important to make the
application early enough in the day for the sulphur
to adhere to the dew on the lower side of the leaves.
F. had been a practical vine-dresser on an Ohio River
hill-side, and had seen the disease there in its worst
form.   So his opinion was worth heeding, that with
thorough and judicious sulphur treatment it may be
conquered.

After another day spent in seeing the cathedral
and visiting the principal cellars in Rheims, I took a
train for the north, and, crossing first the ultimate
boundary of the vine zone and then the Belgian
frontier, was soon far away from gay France, and

among a people whose language is French, but whose temperament is not, for they produce no wine.

I was a few days, too, in Holland, where, though water abounds, it has nevertheless so evil a quality, the people abstain from it almost totally, and drink instead a clear liquid looking much like it, called gin; being, in fact, a wine after Gall's recipe, namely, diluted alcohol flavored with juniper-berries. Notwithstanding, however, their abstinence and their antidote, so humid is the atmosphere, much water enters by the lungs and pores, rendering them lymphatic, hepatic, splenetic, and heavy.

Finding my way out of Holland by the Rhine Valley, I entered Germany, and slept for the first night in the ancient city of Cologne, where the traveler finds a grand cathedral, a sweet perfume, and some uncommonly offensive odors.

CHAPTER XI.

THE RHINE AND JOHANNISBERG.

TO ascend the Rhine, I took passage on a small boat, called a Mississippi steamer because she had a little saloon under the quarter-deck and even with the main-deck, for shelter from the weather and to eat in. In this craft, comfortably enough, we worked up stream, going slowly, as we ought when a double line of curious and beautiful objects is to be seen, enjoyed, and committed to memory.

The vineyards terraced on the stone sides of the steep hills were very well worth seeing—the hard handiwork of a determined people, resolved on getting wine to drink, even if they must smite the rock for it, as Moses did for water. This hardscrabble sort of vine-culture is one of the attractions of the Rhine, and, like Drachenfels, "The Cats," "The Mice," and other well-kept and beautiful ruins, serves to draw yearly crowds of tourists, whose plunder is more valuable to the inhabitants than was formerly that of

the travelers and pilgrims upon whom the nobility
and gentry—original builders of the towers and cas-
tles now in ruins—used to swoop down like eagles of
prey, as they were. The region of steep, terraced hills,
however, is not the *Rhinegau*, where are Rudesheim,
Steinberg, and Johannisberg. The wines wrung with
so much toil from those surfaces of basalt is mostly
sour and hard, like our own Catawba grown on strong
ground, abounding, like it, in tartaric acid, which Liebig says is good for gout, gravel, and stone—by ho-
mœopathic rule of contraries, maybe. It is the base
of much Gallization.

The Grand Turk, returning from a visit to Napoleon, happened to need our comfortable "Mississippi," and so at Coblentz we were put upon a still
smaller craft—an "Ohio," perhaps—in whose close
hold the entire company of passengers had to pack
themselves for shelter from a hard storm of rain,
and so we went by the celebrated district without
seeing it. I blamed myself, on reaching Mayence,
for not getting off at Coblentz, to come on the day
after, so as to see the Rhinegau in clear weather;
but, had I so planned it, the memorable things which
happened to me on that day—a day to be marked in
my calendar with a white stone—yea, with a pearl—
would have been missed forever.

JOHANNISBERG.

The cliffs of basalt that close in on the Rhine as you pass Coblentz in going up stream, and which, according to a habit that hardest of rocks has for always reposing on the softest, lies on a bed of sandstone, give way after you pass Bingen, allowing the sandstone to appear and occupy the surface, which it does in a series of gentle swells and hills. The course of the river is here from east to west, which brings the right bank to face directly south. The zone called the Rhinegau extends from Wallauf, a little below Mayence, down to Rudesheim, a distance of 12 or 15 miles, with a breadth of three or four.

The most conspicuous of its hills is one of mound-like shape and individuality, on whose southern exposure are some fifty or sixty acres in vines, while the top is crowned by the castle or palace of Johannisberg, which is, however, no great things as either palace or castle—no great things, I mean, considering how precious is the ground it stands on.

Every body has heard of it, and knows how it came to the great Metternich on the downfall of the first Napoleonic empire, and is now owned by his descendant, the present Prince Metternich, and Austrian embassador at the court of the Tuileries.

The claims of its rivals to the contrary notwithstanding, there is no doubt Johannisberg is the most
famous vineyard in the world—a very Mecca or Je-
rusalem to the vine-grower or wine-drinker on his
pilgrimage—and pleasanter to visit, I should think,
than either of those musty, holy cities of the past.
The day after the rain-storm was all could be wished,
and I devoted it to the Rhinegau and Johannisberg.
Going on board a down-river steamer, from her deck
I enjoyed a survey of the entire land of promise from
Wallauf to Rudesheim, and even unto Bingen, for
the boat did not stop at Rudesheim.

Crossing to the Rudesheim side in a small boat, I
found myself still about a mile from the town, which
distance I must go afoot. From the uneven shore-
path I strayed upon the well-graveled track of a railway that conveniently led in the direction I was going, but a rough official drove me off with shouts
and gestures. In Germany grown people are watched as tenderly as we do children in America, and
every convenient place for getting crushed, or drown-
ed, or breaking your neck is as thoroughly policed as
if one had not the right to quit when he pleases a
world which wise men have long ago pronounced a
failure.

A German professor remarked to a friend of mine

that he had heard our American police allowed any
one who fancied it to take an unlicensed leap over
Niagara Falls! It is true they do, and Blondin and
Sam Patch are but living and dying examples of
what free institutions will do for a man.

At the railway station I found a hack, which, getting in, I told the driver, "Schloss Johannisberg,"
which was all I knew how to tell him, seeing I spoke
no German. A drive of four or five miles, the last
of it up a gentle rise, overcoming a perpendicular
height of about three hundred feet, I should think,
brought me round to a little village at the back of
the castle, and then through a gateway into the courtyard. If I had looked at the castle instead of the
vines on the way up, I should have noticed, what I
did not, the dark Austrian flag flying from the staff
on the roof. I saw it only after the carriage had
stopped before the principal entrance, and the presence of several servants round the door, the yard
fresh covered with yellow sand, and green-house
plants ranged along the walls, made me look about.
The family were at the castle.

A servant came to the carriage who understood no
French. He went and brought another who understood very little, for, after receiving my card, with
request that he take it to the director, M. Herz-

mansky, whose name, furnished by the waiter at the hotel in Mayence, was absolutely the only clew I possessed to Johannisberg, he went in, but returned in a minute to ask me to alight and enter, while he went to find the princess, who was walking on the terrace.

"Madame the Princess! Pardon me, I only want to see M. Herzmansky on business." Had I told this to the driver when I took the carriage, he would have stopped at the door of the business-office. But, when M. Herzmansky came, the little difficulty of language again occurred. He spoke no French. It was while I was endeavoring, by help of the servant, to explain why I asked of his courtesy admission to whatever it was proper I should see, that a gentleman approached, whose magnificent appearance, more than the general lifting of hats, showed him to be of some distinction. It was Count Edmund Zechy, of Hungary. By his politeness, I was very soon put in proper relation with the director, who, conducting me to his office, asked me to be seated, while his son, just returned from the University, was sent for to be our translator.

His first effort was the doing into French a message from the princess that at four o'clock she would take coffee in the cellar with some friends, and inviting me to be present. To take coffee with a princess would be something; to take coffee with the Princess Metternich was a great deal. To take that coffee with such a princess in the grottoes of Johannisberg was more of good fortune than often falls to one's lot of a summer's day. I accepted; who would have declined?

This was about two o'clock, so there was still abundant time for a walk through the vineyards. Yet there was no more than time, since the good Herzmansky was so pains-taking an instructor, and his replies so full, and he made the tour about the hill so extensive, that when it was ended four o'clock had come.

Those who may be familiar with the strongly-colored red sandstone or sand-shale earth seen about Newark and elsewhere in New Jersey, or with the soils of similar appearance and constitution near New Haven and at other places in Connecticut, would recognize in that of Johannisberg an old acquaintance. The entire hill was of the same; nor did I see any other in any of the Rhinegau vineyards, though I was told there was some slate in places. As the deep color could be due to nothing else than iron, that element must have been present in good large proportion; considering which, I recalled that the Médoc soil abounded in iron, that it was even stronger on the Côte d'or, and was not wanting in Champagne. I recalled also that the places in New Jersey and Connecticut where the like appearance is seen are noted for good fruit and sweet vegetables. It was not, as I should think, by any means a rich ground, and the grain of it was coarse.

Pausing at the steepest part of the hill, which is immediately below the castle, M. Herzmansky called my attention to it as being the place from which the best wine came. There was nothing to distinguish it from the neighboring parts except its steepness, which was so great I inquired why its soil seemed so little washed away after so heavy a rain as had just fallen? He replied that it was because of deep and frequent tillage; but that would not have saved it, I am sure, were it not for its coarse, and loose, and maybe shaly texture. He added that, to compensate for what washing did occur, he hauled on to the higher parts, from time to time, good earth from the north side of the hill, which, slowly descending under the action of rains, not only replenished the substance of the ground, but improved its quality.

The two-forked hoe the laborers were using I noticed to be one half longer in the bit than any I had ever before seen, and that such a tool could be sunk

H

up to the hub at each blow, as I saw the stout-armed fellows doing, showed the ground to be of extraordinary lightness. The hoeing is done four times a year.

Stakes were the only supports. Trellis might be used to advantage, no doubt; but, as with other kings and rulers, the *rôle* of the Johannisberg vines is to be conservative, and let well enough carefully alone. When I asked M. H. if he had heard of M. Liebig's recommendation to ferment wine, like Bavarian beer (our lager), in open vats and at a low temperature, he answered that Liebig had spent some weeks at Johannisberg, and had, while there, suggested that very thing, which he (M. H.) thought might do very well, but he also thought experiments should first be tried on less valuable grape-must than what his vines gave.

The vines were of good size, and spaced, I believe, like nearly all I saw in Germany, about three feet apart. Care had been taken not to overcharge them with fruit.

Surprised to see a square piece of what seemed as good vine-ground as the rest covered with only a crop of clover, I asked the reason. It seemed the piece in question had lately accomplished its fiftieth year of grape-bearing, and, according to the traditional usages of the place, had been liberated from its vines, which

had been thoroughly extirpated, and was now in clover, keeping its jubilee of three years, at the end of
which time it would be planted again, after receiving
an overturning to the depth of three feet.

Vintage, which is delayed as long as possible, is
conducted somewhat as in the Sauterne district, the
fruit being gathered as it ripens, and selected berry
by berry, so that the ground is three and four times
gone over before all is done, which carries them often
well into November. These late vintages and successive gatherings are, of late years, the general usage
throughout Germany. Dry weather is thought essential for the gathering, and, before pressing, the grapes
are often spread out to dry during from twelve to
twenty-four hours. I am pretty sure they told me
they pressed without allowing any previous fermentation on the skin.

So rich in sugar is Johannisberger that its final
maturity and clarification is the work of about seven
years, during which, from time to time, it repeats, as
it were, its second fermentation. All great wine ripens slowly, I have noticed, and from the same cause,
I suppose. The casks hold, as I should guess, near
200 gallons. One of them, called "the bride of the
cellar," contains the best wine in store, and is kept in
reserve till its mate in excellence is found, to which

it then gives place. The Germans are full of such pretty fancies.

From the character of the soil and appearance of the vines, I judged manure was used, and asked if it were so. M. Herzmansky replied that it was, and after we had done the vines he showed me a large stable of admirable arrangement, where a number of cows were kept and "soiled." Others, to the number of forty in all, were let out to the work-people on condition to be similarly kept, and the manure delivered as rent for their use. No other kind than this was used, and of it a good deal was applied. M. H. insisted that no injury to quality resulted, and even thought it improved the bouquet. It may be so; judges often differ. Many men have many minds. Circumstances alter cases. What's one man's meat is another man's poison. The monks of old Vougeot let their vines stand 500 years, while those of old Johannisberg uprooted theirs every fifty years.

The monks once owned Johannisberg, then? To be sure they did—the monks of St. John—only another coincidence, miracle, or special providence. Should all the good things in this world go to the share of the sinners or of the saints, I would like to know?

But then the saints drank up *all* the good wine.

The poor sinners, destined to so hard a fate in the end, needed some little consolation in the way of present creature comforts.

But the monks had no wives or sweethearts.

Poor fellows—more to be pitied than the poor sinners—it was right to let them have the wine!

"Here is an American vine," said young M. Herzmansky as we left the stables, and, sure enough, there was an Isabella, trained to a wall, and growing well. They surprised me by remarking that its fruit had too rank a flavor. Returning to the office, the director showed me his labels, and, at my request, gave me a sample of each kind, signed with his name, as is that of every bottle sent from the cellar, and also a list of prices. Here is a copy of it, the prices being per bottle:

				Flor.	Kreut.
1857, Cabinet wine			seal	4	
"	"	"	"	2	30
1858,	"	"	"	9	
"	"	"	"	7	
"	"	"	"	4	
"	"	"	"	2	30
1859,	"	"	"	20	
1862,	"	"	"	14	
"	"	"	"	10	
"	"	"	"	7	
"	"	"	"	4	

The florin is equal to forty cents, and sixty kreutzers make a florin. It seems the administration, in which M. Ebenhöch is associated with M. Herzmansky, receive and fill orders addressed directly to them at Schloss Johannisberg, authenticating every bottle, as I said, with the signature of the director, M. Joh. Herzmansky. I need not say the chances are poor for obtaining the genuine thing through the ordinary channels of commerce. It will be seen that, except for the very highest grades, which are only to be tasted on great occasions, the prices are not by any means extravagant.

Many years ago an old man of the neighborhood came to the castle and offered his services, so goes the story, without compensation, to oversee the labor in the vineyards. His offer was accepted, and for a long time he performed the duties he had assumed from love of them. It was worth something to be chief vine-dresser on the slopes of Olympus.

Four o'clock had come, and the gods were assembling—that is to say, a party of ladies and gentlemen were descending the steps of a door at the side of the castle, and were about entering that of the cellars which opened close by it. One of them, a lady of the Hungarian type of face, cloaked and hooded for

the descent into the lower regions, met me as I drew
near, and addressed me in perfect English.

I told her why I had come to Johannisberg, adding that, had I known, however, of her being there, I
should not have taken the liberty.

"But why not?" she said. "I'm sure I feel very
much obliged to you for taking the trouble."

Which made me feel I was no intruder.

The cellars were right under the schloss. As we
entered, I could perceive that it was wide and high,
arched overhead, and remarkably dry and clean.
Numerous candles lighted it brilliantly. At the upper end was a sort of stage scene formed of green-
house trees, shrubs, and flowers, and other ornaments,
having quite a pretty effect.

It seemed it was a wine-tasting party, composed of
people assembled from the neighborhood, and some
friends then on a visit to the princess. And on
that day of all others, when wine was to rain down,
it was my fortune to be there with my goblet held
up!

After making a tour of the alleys, we reached a
place where a table was spread and chairs arranged,
around which the company seated themselves. After
an explanation from M. Herzmansky that, though her
highness had ordered coffee, yet, as it was inconsist-

ent with that perfect equilibrium of the senses and freshness of the organs of taste required where his nectar was to be received, he had taken the liberty of suppressing it, and would at once proceed to place before us samples of different vintages in the order of their merit, beginning with one of the lower grades (all which was explained to me by Count Zechy, next whom I sat), the wine was then served in green Hock glasses of good capacity. The first, and least good of the eight or ten different sorts, was much the best white wine I had ever tasted. It was followed by a better, and then a better, and then a better still, mounting in scale of graduated excellence—the enthusiasm rising, too, in like measure—until the best was reached in form of a cask twenty-one years old—just come to its majority, in the sipping of which one could only exclaim, Wonderful! wonderful! "The bride of the cellar" was yet to come; and she came, radiant and delightful as star-crowned Ariadne, bride of Bacchus.

Is it expected I should describe these upper Johannisbergers? Epithets, comparatives, and superlatives gave out in exhaustion a long way down the ladder. "Richness," "fineness," "softness," "body," "vinosity," "flavor," "bouquet," terms of commerce and of table small-talk, all apply to wine, and are limited in meaning by the finite qualities of their subject. But I am writing of JOHANNISBERGER.

Shall I talk poetry to it, and try to compass its excellence with figures and flowers of speech? Or shall I quote, where I dare not originate, and make the Great Master of the lyre speak for me? Let us see. He has done his tersest and best when he tells us Ulysses took with him to the isle of the Cyclops wine that was

"Rich, unadulterate, and fit for gods to drink."

But would the gods themselves be fit
For drink so rich and pure as it?

"Look there!" said Count Zechy. I looked up, and saw the scene I have named, at the other end of the grotto, slowly rise in the air, floating among clouds of violet and rose, through which shone rays of every iris tint—the scene itself changing, as it moved, into a celestial landscape. For a long while it remained in view. Now it approached and now receded, all the time rising, and so, gently coming and going, and slowly mounting upward, it passed away.

Was it "the bride of the cellar" had made us see this beautiful thing?

No; it was M. Herzmansky burning Bengal lights among the green-house plants.

As, in the course of the tasting, I had taken to the

H 2

quantity of two or three glasses, or maybe more, it is proper I should say something of the effects of those high Johannisbergers.

To begin, the first effect was the same as poor Proserpine felt when, having tasted the juice of pomegranates grown in Pluto's dominions, she wanted to live there through all eternity. So felt I, in those lower regions of Johannisberg. After partaking of the charmed fruit, I was spell-bound, and tore myself not easily away.

Coffee excites the brain, and tea the liver, with an unbalanced action. Hasheesh and opium are liars which cheat the senses with unreal objects, born of the vapors of the brain they disorder. But wine acts honestly on the real, and, while exalting the action of every part within us, and the effect of every object without us, works only with the actual and the true. Accordingly, as I drove down hill, the working of the spell that lay in the kiss upon my lips the cellar's bride gave glorified all the wide Rhineland, already glorious with the sunset, gving to the waters a brighter sparkle, and deepening the purple of the shadowy hills. By the same power, every event since the morning came over the mind again and again, with repeated titillations of memory's chord, multiplying that one delightful day into many. I was contented

with all things and all events, for all had treated me
well. The river and the hills had pleased me; the
air and skies had been amiable; the people I had
met had been polite and kind; and I realized most
vividly—what was true most really—that the heav-
ens were good, and the earth was good, and the fruits
of it passing description—that this was the best possible of worlds, and the queen of its queens the
Princess Metternich.

CHAPTER XII.

SWISS VINEYARDS.

A TRANSIT across Germany to Salzburg by rail,
and thence through the Alps of the Salzkam-
mergut, a two days' carriage drive, brought me to
Bad - Gastein, without any thing occurring by the
way with a bearing on the object of this book, unless it be that the more I saw of wine-drinking South
Germany, the more was I charmed with the manners
of the people.   In respect to manners, which are the
flower and fragrance of benevolence, the Austrian
Germans are, in my opinion, no way behind the
French.   Indeed, what I saw of Austria made me
more than ever a believer in the theory of Babrius
concerning the influence of wine upon civilization.
Like the French, they have exquisite taste.   When
they shall have become fully possessed of themselves,
as will soon happen, through the development of freedom under the rule of Van Beust, great things may
be expected of the Austrians.

That great man was at Gastein while I was there, and the daily sight of him was good for the eyes. Let me predict he will do more for his people by working on their better nature than his now triumphant rival, Bismarck, ever can for his through appeals to their bad passions, party violence, and the use of force.

But what has all this to do with white wine or red? Very little, I confess.

The tour I afterward took in Switzerland made me, of course, familiar with Swiss vineyards, for every canton but two contains them. The last week of September and the first of October I spent at Glion, *sur Montreux*, in the heart of the vine region of Lake Geneva, the most important of all. In that lovely cove, formed by the head of the lake, there are a hundred hotels and boarding-houses, where invalids resort in the autumn to cure themselves by eating grapes "by the quantity." I tasted often of the best in the neighborhood, and found them neither sweet nor well-flavored.

I have not yet heard of the grape-cure in America; but Catawbas *will* cure, though. They will cure, for instance, summer complaint and autumnal dysentery, and, if Liebig is right, will also eradicate gout and calculus—all by virtue of the tartar they contain, to say nothing of what their other constituents may be good for.

I could account for the insipidity of the Vaudois grapes, as well as the poor reputation of the wine made from them, when I learned that by dint of heavy manuring the vine-dressers of the Canton Vaud had carried the yearly average yield per acre up to about a thousand gallons. They are most thoroughgoing cultivators, and no doubt understand their own interest in pushing their vines as they do. At the same time, they produce two or three qualities of good repute. The soil about Montreux is basaltic, and so is that of a large portion of the region about the lake. On the way northward from there, along the western border of Switzerland, where the vine abounds, the formation is almost wholly of a soft, light gray sandstone. Here we have basalt and sandstone in abundance, but no very good wine. But the sandstone is not red, and is probably without admixture of iron; nor could I see that it was shaly at all.

CHAPTER XIII.

DURKHEIM.

ON my way to a second visit among the vineyards of the Rhine, I found myself, on the 12th
of October, at Freiburg, in Breisgau, in the upper
valley of that river. There is a great hill back of
the town where wine of some repute is grown, but
there the vintage had not yet begun. In the low
grounds it was well advanced. The first sight I met,
in walking among these last, was a boy standing, or
rather dancing, in a tubful of grapes, which rested
on an upright cask with the head out, into which the
juice flowed from numerous auger-holes in the bottom of the tub. There was also a trap in the same
bottom, which was lifted from time to time to let
the crushed remainder of skins, seeds, stems, etc.,
fall through. This seemed to be the fashion of the
neighborhood, and an old fashion it was too, having
been brought from Persia, whence came to Europe
(as is generally believed) the grape and the art of
making wine, and where, to this day, they express

the juice in the above manner. The grapes were white. They were to be allowed a twenty-four hours' fermentation on the skins after crushing, in order to improve the bouquet, it is said, and then put under the press.

Keeping down the valley, I arrived at Durkheim, in Rhenish Bavaria, a central point in the important vine region which includes the towns of Forst and Deidesheim. Though it was already the middle of October, vintage had not yet begun, and I learned, what I ought to have known before, that of late years they let the fruit hang on the stem till almost ready to fall. The hotel was a coarse old concern, whose rooms were either comfortless from absence of fire, or uncomfortable from the presence of stoves and the smoke-nuisance. Many of the guests were come to eat grapes as medicine, like those I had just left on Lake Geneva. But the Durkheim medicine was not by any means so bad to take as was that of the other place. In fact, the grapes were excellent, and by far the best I found in Europe.

I suppose Brillat Savarin, were he alive, would object to take this grape-medicine; for once, when some one offered him grapes to eat, he declined, saying,

"Je ne prends pas mon vin en pillules" (I don't take my wine in pills).

There are hills in the neighborhood of Durkheim, but the soil to which the characteristic quality and value of its wines and those of Forst and Deidesheim are due is found on the surface of a wide and level area of gravelly deposit, so permeable, and consequently so poor, that a soil has to be made for it, and kept continually renewed, by hauling upon it basaltic earth and clay, together with large quantities of cow manure, which last the first two serve to retain, so it shall not be washed too freely through the sieve-like foundation beneath.

I found no difficulty in obtaining the guidance of a young gentleman, who was the son of a large proproprietor, and who was good enough to devote the whole day to me. Beyond question, the Rhenish Bavarians are the first vine-dressers of the world. Their vines are wide-spaced enough for plowing, yet all is done by hand; and how often in a season, think you, do those sturdy fellows stir the soil? Nine and ten times! On each acre they yearly bestow a hundred and forty days of hard labor. From distances of many miles they haul basaltic earth in such quantities as in time visibly to elevate the surface of the vineyards above surrounding fields. To this earth they seemed to attach more value than I did, after what I had seen on Lake Geneva. Its good

properties seemed only to be manifest when it was
spread on other ground, and may have consisted
merely in the attraction its dark color has for the
sun's rays and its retentiveness of manure.

The wines thus produced are uncommonly fine,
and mostly rich in bouquet — very different from
what could be expected with such heavy manuring
bestowed on a clayey or even a basalt soil.

The vines are of good stature, and trained on wire
trellis.   The oïdium is not an uncommon visitor, but
is successfully met with sulphur treatment.   Observing a sprinkling of whitewash on the vine-leaves
along the outer border, and recalling the verdigris
sprinkling noticed in Médoc, I asked the reason of it,
but could get none better than that it rendered the
ripening fruit unattractive—not to birds, but to boys.

In the cellar of the father of young —— I saw
the casks had received a good coat of coal tar, the
object being, they said, to retain the freshness of the
wine as early bottling will do.   I think it was but
an experiment, and not an established usage.   It
would certainly need to be very carefully done, and
the odor should be well dried away before putting
the casks in use.

Quitting the Rhine by the valley of the Maine, I
had a glimpse of the vineyards producing the cele-

brated Stein wine, which they put up in big-bellied bottles that look generous and honest, and are so, certainly, in comparison with the slim, long-necked flasks we cheat with at home, and yet they are themselves very short-comers in contrast with the full quart flasks sometimes seen in France, which last, I hope, will long be preserved as monuments of departed honesty.

Bow pruning seems to be a favorite in Rhenish Germany. Probably the strong manuring the vines receive in that country enables them to bear what it is insisted would be ruinous in France. And it may be that manure can be more freely used, without injury to the vine, on the extremely porous, gravelly plains, or well-drained, terraced mountain sides of the Lower Rhine, than on French soils. Guyot, we have seen, accompanies his recommendation of long pruning with the requirement of high manuring.

Undoubtedly the tendency in our day is to cultivate for quantity rather than quality, just as in cookery it is to sacrifice taste to convenience or economy. We invent no new dishes, but only quick, cheap, and easy ways of spoiling old ones. Modern improvements in the kitchen consist in neutralizing the flavor of our vegetables by boiling them with soda ; raising bread with chemicals instead of yeast, and baking it

with a cast-iron heat instead of the gentler radiation
of a clay surface; and as to joints and poultry, in
thrusting them into the oven with a dripping-pan
full of water, and there " soddening with water,"
against the express prohibition of the Bible, instead
of "roasting with fire," as it expressly commands.
And Science and Invention are spoiling our dinners
for us as well as our wines.

CHAPTER XIV.

VIENNA, AND BEER, AND TOKAY.

VIENNA, where I remained three weeks, beats the world, and even Paris, for bread, beer, and boots. I found the best beer, not where the Strausches perform, but in a cellar near the Opera-House, where, thirty feet under ground, a deal of deep drinking is done, and the new luxury of political discussion enjoyed.

If, as wise Babrius tells us, every drink has its peculiar effect, that of beer must be metaphysical, disputatious, over-refining. It must be so, for something has split German philosophy into half hairs, and German nationality into a hundred or more states. It could not have been wine, which is a unifier; so it must have been beer. All the world knows that German freedom and unity could have been secured in 1848 if the Frankfort Convention had not lost its head in a fog of abstractions, and spent six months over a Bill of Rights preliminary to a Constitution, allowing the kings to take courage, gather strength, and drive the abstractionists out of the windows by the practical pricking of bayonets behind, in the thirteenth month of their incubation.

A few years ago, while yet the Glossners brewed beer as good as the best at their cellar in Cincinnati, I happened to be conversing there with some intelligent Germans—a judge, a lawyer, and an editor—about free trade and protection. They talked rationally enough till the question arose, in the course of the discussion—but how, I can't imagine—whether light was positive and darkness negative, or whether it was the reverse. With an apology to me, they soon plunged into very excited German talk, replenishing their glasses frequently. After each had drunk his quart, I rose to take leave, but they begged me to remain, promising they would soon conclude their metaphysics, and return to common sense and the English language.

" But what have light and darkness to do with dry goods, hardware, and groceries, and their importation ?" I asked.

They assured me they could not possibly get on with the main question until the other was settled, and at it again they went. I left them at the end of another quart, wrangling as positively and negatively as in the beginning.

While at Vienna I had an interview with Count Henri Zechy (cousin of the Count Zechy I met at Johannisberg), to whom I had been recommended to apply as one of the best-informed of the Hungarian wine - growers, and whom, when he called, I recognized as one of the Austrian commissioners at the Paris Exhibition, where he was, I think, chief of a group. He resides near Ödenburg, in the northern part of Hungary, which is in the region where Tokay wine is made.

He told me they made all their good wines from soils good for nothing else. Little manure is used except in layering, which they practice to supply failures. They think it injures the quality. On steep hill-sides they heap up ridges and mounds to retard the running away of the rains, thus retaining enough to supply the roots, and, at the same time, preventing loss by washing. Cultivation, always by hand, is done three times in a season. In the Tokay region and other northern vine districts they cover up in winter by laying the vines on the ground, and then spreading straw over them, and afterward earth over the straw.

No oïdium is known in Hungary, unless we may recognize as such something believed there to be the sting of an insect, which appears on the young grapes when no larger than shot, arresting their growth, and causing them to fall off.

The sweet wine known as Tokay can only be made when the season will permit more or less of the fruit to dry while yet hanging on the vines. If, during October, there are white frosts in the mornings, followed by clear and dry days, by the end of that month a portion of the berries will be found to have shriveled almost like a Malaga raisin, and become blue. In such case, the berries thus dried are carefully picked by themselves and reserved for a special treatment, the remainder being afterward gathered and pressed in the usual way.

If the selected grapes are in sufficient quantity to express any portion of juice by their mere weight when flung into a vat, such juice, called " essence," is kept by itself, and, when sold separately, brings extravagant prices. After this, or without it, if the quantity will not suffice for a flow of " essence," the dried grapes are put into sacks, and trodden with feet until reduced to a perfect marmalade. The juice flowing in the course of this operation being, like the " essence," kept apart, becomes Tokay wine of the first class. The " marmalade" remaining in the sacks is then pressed, to yield a Tokay of the second class. These products, namely, the essence, and first and second class wines, are afterward mixed with what is made from the fruit which did not dry, in the proportion of 1 to 10, 2 to 10, 3 to 10, or 4 to 10, and the qualities of sweet wine thus compounded are named, and rank accordingly.

If the weather has not been propitious, of course there can be no sweet wine made. Birds are very troublesome, as might be expected, where such sweet fruit hangs out of doors so long, and have to be abated with guns, or frightened away with machines contrived to make a noise.

Not over one per cent. of the wines of Hungary are sweet. To eke out this small supply, they make, at Odenburg, an imitation Tokay, by mashing up Greek currants and Malaga raisins, which is apt, however, to spoil within a year or two.

A simpler way of making the Tokay would be to gather both the dried and undried fruit without culling, and press all together, but this would not give such regular and certain results. Besides, without a thorough mashing in the bag, but little juice could be obtained from grapes so dry, as, mingled with the others, they would almost escape pressure.

Count Zechy told me the average of alcohol in Hungarian wines was about ten per cent. He remarked that they contained a perceptible proportion

I

of phosphorus. I have heard the Burgundy wines also
contain it, and to its presence is attributed, by some,
the greater "headiness" of Burgundy as compared
with Bordeaux, phosphorus—as it is known, having a
close similarity to the substance of the brain. The
grape soils of Hungary are of different kinds—limestone, chalk, sandstone, and basalt. I should call the
weather Count Zechy described as being good for
making sweet Tokay excellent for ripening persim-
mons and pawpaws, and attribute as much virtue to
the morning frosts as to the dry days. We have in
many parts of America so much of this autumnal
weather, I dare say we might try the making of
sweet wines with good hopes of success. "Straw
wines," as they are called, are made by drying the
gathered grapes for several weeks under cover, usually on straw, whence comes the name. I remember
tasting some Catawba which had been made in this
way by a gentleman who lived near Cincinnati, and
that it was quite strong and Madeira-like, though not
very sweet, and had kept well without having ever
been below ground. But if it be possible for some
of us in America to emulate the Tokay, it is certain
none in Europe outside of the chosen region can do
so, as many failures have proved. Something of the
sort is made, however, at the Cape of Good Hope.

The American palate would take kindly to a good sweet wine. In Champagne they know this, and sweeten wines for our market as they do no others except what are sent to Russia, remarking of the two great peoples in question, " Barbarians love sugar."

CHAPTER XV.

ITALY.

OVER the Somering, and into Italy, Venice, Florence, Rome, Naples, where Vesuvius, in a ferment, overflows with her ruddy and well-sulphured wines, and we come to Sorrento, on the south side of the Bay of Naples, even unto Villa Rispoli and its orange-groves, and there rest.

I am sure the "Lachryma Christi" they gave at the villa was not genuine, for it was no great things, which every body says the true wine is.

I had heard in France, and also encountered in my reading, so much in disparagement of Italian modes of vine-culture, that I had really paid little attention to it until what I saw during a three weeks' sojourn at Sorrento made me think somewhat on the subject; and reflecting that even in the lazy province of Naples all the people could not be drones, much less fools, and that men whose palates were discriminating and whose fingers were cunning in cookery must know how to make wine, I made a few turns in the vineyards of the neighborhood, and afterward spent an evening in conversing on the subject with my host, whom I invited to my room for that purpose. His description of the modes of culture in vogue about the Bay — of gathering, fermenting, and keeping — though not applicable, I think, in our own country, and so not worth detailing here, seemed reasonable and proper enough.

For the custom of high training, either on trees or trellis, he could give me no reason; but he might have given this—that in so hot a climate, and on such rich, dark, volcanic soil, grapes grown on low-trained vines would either dry up to a raisin or fall off without ripening. Notwithstanding all said against their methods, I think the Italians know what they are about; and as to their keeping the wine in demijohns and flasks, corked with olive oil floating in the neck, there may be good reasons for that too. An exceptional soil and climate may very well produce exceptional wines, needing exceptional treatment.

Going among the vineyards near Sorrento, and noticing the vine-dressers at their work, I could see that they were careful and thorough in all they did. The trellis were very high, and were made by driving tall stakes in the ground, and tying to them cross-pieces of cane. Certainly the form was not contrived to favor laziness.

The ground was free of weeds. The space for admitting sun and air was about what it should be. The pruning was being carefully done, according to their system, whatever it was; and by what right could I tell them it was all wrong?

They never vintage until after the autumnal rains have come to swell the fruit, giving for reason that the wine would spoil if they did. This certainly would not be a reason in France, and yet it may be a good one in Southern Italy, for all that.

I remember meeting, while in Naples, a Polish lady who owned vineyards near Florence, and who told me she was determined to adopt the French system of training and pruning, notwithstanding her Italian neighbors warned her it would never do. I remember, too, that a few days afterward, another lady told me her father, a Frenchman, had, in fact, tried the experiment on an Italian vineyard, with the loss of his vines as the result. No one would contend, I think, that the method, so successful in Champagne, of crowding 25,000 plants within the compass of an acre, could possibly do on the warm, rich soil of Naples.

I had heard of the wine of the island of Capri

as excellent, and of the blue grotto there as beautiful, so sailed over to it one day. But the high sea
kept us out of the grotto, and the wine was not good,
but bad, from the effects of the sulphur cure, they
said. There is a remedy for the bad effect in question, which is itself a sulphur cure, and which may
not yet be as well known in Italy as it is in France.
At any rate, the "lazy Neapolitans" have done what
no Americans have yet had the patience to do—they
have cured the oïdium.

We remained two days at Capri, where we enjoyed the society of some English artists, who had
gone there to paint from living models, whose perfection of form was beyond any thing to be found
in England—at reasonable rates. The hotel had been
a monastery once, which accounts for the excellence
of the wine formerly grown on the island. The progress of the age has expelled the monks, and the progress of the oïdium has spoiled the wine—a judgment,
no doubt.

Here comes the question, Were those ages during
which monasteries flourished the most, dark ages because good and learned men secluded themselves
from the world, hiding in cloisters that wisdom which
should have enlightened it, and in cellars the bottled-
up quintessence of civilization which should have re-

fined it ? or was it because of the dark and evil times mild and good men were forced to shelter themselves in cells, and keep their liquor under lock and key?

However it may be, those cowled, good fellows had a good time of it. Safe and sure, tranquil and content, they fermented good wine, and brewed bad metaphysics; the wine they kept to themselves, but the metaphysics they let loose in the world to bother mankind, and deepen still more the darkness of the ages.

Soon after leaving Capri, I spent several weeks at the neighboring island of Ischia. At both places the vine-culture was substantially the same as at Sorrento, and at both places they cured oïdium unfailingly.

At Rome, a gentleman, whose father was a large owner of vineyards, gave me several kinds of wine to taste. All were decidedly pleasant, and the Aliatico delicious, but most of them having any age were more or less pricked. This gentleman, in giving some details of their modes of cultivation in the Roman territory, remarked that, for heavy work— trenching three feet deep, for instance—the most reliable laborers were from the province of Naples, describing them as strong, willing, and of excellent conduct.

From Rome I traveled northward into Tuscany,

where cultivation in all branches is thorough, systematic, and careful, and there I found no vines trained
either on stake or trellis; all were clambering in
tree-tops. Twenty-five feet was usually the distance
between the trees on level ground, and fifteen feet
on hills. Two or three vines were planted at the foot
of each tree. This system is not confined to Italy
alone; it is practiced in portions of France also. In
the north of Italy it is common to prune the trees,
so as to let in air and sunshine, while in parts of
the south care is taken to keep them shaded. We
often hear of vines grown upon trees in our own
country, which, for some reason, escape disease, and
from such facts an argument is drawn in favor of
long and high training; but the immunity is probably due to the shelter from radiation which the foliage of the tree affords. M. Du Brieuil tells us vines
trained upon trees in France suffer more than those
on stakes. I learned the same thing to be true of
trellis-grown vines in Burgundy. We know that in
Italy neither trees nor trellis avail aught, and we
shall find that in Southern France the lowest vines
are least afflicted, and the highest suffer the most.

I left Italy by a wondrous road which skirts the
Maritime Alps on one hand, and the Mediterranean
on the other, and is called "Riviera" at one end, and

"Cornice" at the other, traveling in a carriage hired for the whole journey at Spezzia, where it begins.

It is a journey of six days, but so varied and so beautiful in all its ups and downs, ins and outs, that when it ends one is tempted to turn about and go back over it again. Now we descended to the very edge of the sea, and traveled for long reaches on its pebbly or sandy beach; now, mounting high, were whirled at a gallop along the verge of a precipice; now we rounded a rocky cape, on whose bleak sides no plant could stand, and now turned into a cove luxuriant with olive-trees and vines. Those who love the blue Mediterranean may thus, curving about her shores, embrace her, as it were, in a delightful week of prolonged leave-taking, and part from her at last more in love than ever. For nearly the whole distance the abrupt sides of the mountains were terraced with walls of stone, almost from foot to crown, and the soil thus secured planted in vineyards and olive-orchards. Much as we praise the Hollanders for building the dikes which keep back the sea from coming in upon their lands, the Italians deserve scarcely less credit for those dikes of stone which keep theirs from tumbling down into the water. As to the terrace-work of the Rhinelanders, it is as nothing in comparison.

CHAPTER XVI.

THE SOUTH OF FRANCE.

AT Nice I entered on the great vine-country of Southern France, where an enormous quantity of common, and a moderate quantity of superior wines are produced. In this region — namely, at Nice, Nismes, Montpellier, Cette, and other places—I remained about six weeks, with two subjects of inquiry in view—one, the vine disease, and the other, training *en souche*. What I learned on these points elsewhere I have mainly reserved for this place in my book, because in Southern France it is that the disease has been the most virulent and been most triumphantly subdued, and there it is that from time immemorial all the vines have been kept *en souche basse* (on low stocks).

The vine region in question extends from Nice in the east to Leucate in the west, and lies mostly between the 43d and 44th degrees of latitude, though extending as far down as below the 43d degree on the

western wing, and, where the valley of the Rhone is
included, going nearly as far northward as the 45th.
It reaches from the Alps to the Pyrenees, and includes the entire French Mediterranean coast. It is
sheltered from the winds of winter by a line of
mountains that bound its whole northern border,
and from whose bases the whole surface slopes gradually down to the shores of the sea. Composed of
portions of ancient Languedoc and Provence, it includes the present departments of Drôme, Ardèche,
Vaucluse, Basses-Alps, Var, Bouches-du-Rhone, Gard,
Herault, Aude, and Pyrenées-Orientales. Of its entire tillable surface, fully one fourth is in vines—
namely, a million and a half of acres, and the culture is continually extending.

The formation is generally limestone. The soils
are various. On the poor slopes at the base of the
mountains very superior wine is grown. Below
them, at different stages of elevation, but mostly of
level or slightly-inclined surface, are strong, but not
over-rich soils, clayey, limy, and sandy in different
proportions, capable of yielding large crops of strong,
sound wine, which sells, when new, at from ten to
twenty-five cents a gallon. Here and there on the
level ground are found pebbly deposits whose product, like that of the poor hill-sides, is of a high or-

der. Finally, the rich alluvial borders of the rivers have been known to produce, per acre, in a favorable season, as much as 4000 gallons of weak wine, containing only six per cent. of alcohol, formerly destined for the still, but of late years used to compound with other sorts in making cheap wines of commerce.

Observations taken at Montpellier show the climate of the region I am trying to describe to be marked by strong peculiarities. The mercury rises above 86° Fahrenheit, on an average, 34 times in a year. There are, in a year, 174 fair days (at Paris there are only 56). The mean number of rainy days in a year is 81. The yearly rain-fall averages 924 millimetres, about 36 inches. Rain often comes in torrents, as it does in America, but does not any where else in France. But little rain usually falls between the middle of June and the middle of October, an advantage which is somewhat compensated for by the heavy rains in the last half of October.

For the reason that the large average rain-fall of 36 inches is poured down in comparatively few rainy days, and the farther one that the prevailing winds are mostly violent and drying, the climate is a very dry one.

Although the mountains on the north are a shelter against cold winds coming from beyond them.

they themselves give forth the frequent, sudden, violent, persistent, and biting cold "*mistrall*."

Thus we find in the south of France intense summer heat, sudden and rude changes of temperature, high winds, heavy rains, and great dryness of amosphere, making up a climate' very much resembling our own, and very little resembling that of any other part of France or of Germany.

When I first visited Languedoc, all its broad fields were in the plenitude of their autumnal array, the vines wearing their green, their purple, and their pearls displayed on outspreading and low-trailing branches, as if each one were a belle or a bay-tree. The next time I saw them, which was in March, and after winter-pruning, nothing met the eye but little, low brown stocks ten inches high, with all their branches cropped to two or three inches. Crinoline had given place to fact.

THE OÏDIUM.

Maybe many of my readers will think the pages given to this subject contain nothing important for them to know. Let them not be too sure of this. There may be regions in America, as there are in Europe, where the scourge has not yet come, and may never come, but such will be exceptional, and we can not yet know them. Cold latitudes are not propitious to the growth of the parasitic plant we call oïdium,. and accordingly we find the more northern limits of our vine zone have thus far been most free from it. The disease has more than one form, and has been often mistaken for a mere leaf-blight by those who think themselves far beyond reach of the oïdium. New vines are generally strong enough to fling off all ailments which beset them for the first few years after they come into bearing. During those years they will commonly thrive and produce well, so that results obtained from such often lead us into error as to the value of soils, varieties, and modes of cultivation, as Mr. Sanders very well remarks in one of his late publications.

I have three vineyards of Catawbas which came into bearing in 1860, and continued to do well and showed no sign of disease until 1864, when the disease destroyed about one tenth of their fruit. The next year there was a clean sweep, and the next, and next.

As with new plantations, so it is with new varieties. The Norton's Virginia Seedling, the Concord, and the Ives Seedling are the three which have been most confidently relied on and most loudly praised for their invulnerability. My own Nortons, that had remained safe and sound since I first planted them in 1857, had the disease last year (1868), not merely in the form of "gray rot," but also in that of "*red-leaf*," the most terrible of all its manifestations.   In the month of September of that same 1868; I saw, in a four-year old vineyard at M'Arthur, Vinton County, Ohio, as pretty a rotting going on as one who had foretold it would wish to see, while close beside was a vineyard of three-year old plants loaded with fruit and in perfect health.   The same season, the Concords of one of my neighbors on the banks of the Ohio suffered badly and for the first time.   Then, for the Ives, what meeting of the Cincinnati Horticultural Society or Wine-growers' Association has there been since August, 1868, when Mr. Howarth has not risen in his place to declare that it *does* rot, and offer to prove it?   The Dianas, Rogers No. 15, and other Rogers plants, have also given way before the advancing pest, and the Delaware too.

The Catawba, that has been made a scapegoat and abandoned as hopelessly doomed, is, in fact, remarkably hardy in resisting the disease.   This has been repeatedly noted in France, where it is grown experimentally.   On the Lake islands and Lake shore in Ohio it withstood the invasion year after year, and fortunes were made from its fruit before it succumbed. If their Concords or Ives's hold out as long, I shall be surprised. In my opinion, the Catawba is better proof against the attempts of the destroyer than almost any variety we have while, of those whose hardiness so many have been willing to vouch for, the toughest can only hope to be reserved for the honor of being *last* devoured.

It better behooves our vine-dressers to examine into the disease, learn the remedy, and prepare to apply it, than hug themselves in an illusory security, or fly in a panic from one variety to another, or from one place to another. But mildew, rot, and red-leaf —in other words, oïdium—can be cured and kept down. It has been done in Europe, where its march was far more rapid and sweeping than here, and here we can do it too. The evening after I arrived in Nice I bought the pamphlet of Mr. H. Marès, whom I have already mentioned as having received, at the distribution of prizes at the Paris Exhibition, a medal and a smile. I took the brochure home, and did not sleep till I had read it through. Here it is, and the reader must read it too, every word of it, that he may the better understand what is to follow it.

MANUAL

FOR THE

SULPHURING OF DISEASED VINES,

AND

RESULTS.

By H. H. MARÈS, MONTPELLIER.

TABLE OF CONTENTS.

EXPLANATION OF THE FIGURES OF THE PLATE.

Fig. 1. Reproductive spore or germ of the *Oïdium Tuckeri*, largely magnified.

Fig. 2. *Oïdium Tuckeri*, largely magnified. *m, m, m.* Creeping filaments, or mycelium. *c, c.* "*Crampons*" (clamps, anchors) of the mycelium. *t, t, t.* "*Tigelles*," or erect filaments bearing the spores placed end to end. *s, s, s.* Spores on the "*Tigelles.*"

Fig. 3. The *Oïdium Tuckeri* in full vegetation on the skin of a grape, appearing to the naked eye as merely a white efflorescence.

Fig. 4. Spore of the *Oïdium Tuckeri* beginning to germinate.

Fig. 5. Fragment of the skin attacked by *Oïdium*, on which flour of sulphur has been scattered. *f, f, f.* Grains of flour of sulphur.

Fig. 6. *m, m, m.* Fragments of *mycelium* broken and deformed. *s, s, s,* Spores shrunk and distorted. The greater part have disappeared; only a small number are seen.

Fig. 7. Flour of sulphur, magnified.

Fig. 8. Fine *sleet* of sulphur, largely magnified.

Fig. 10. Triturated sulphur reduced to an impalpable powder, magnified to the same degree as the flour of sulphur in Fig. 7. Common triturated sulphur has nearly the same forms, but the fragments are much more voluminous.

Fig. 11. The Vergnes Bellows. *t.* Nozzle by which the air enters and is blown out again, charged with the sulphur-dust. *m.* Wire gauze with large meshes to sift the sulphur. *c.* Cavity of the bellows, serving as reservoir for the sulphur. *b.* Stopper to the orifice in the upper wood, by which the sulphur is introduced.

Fig. 12. Box with a tuft.

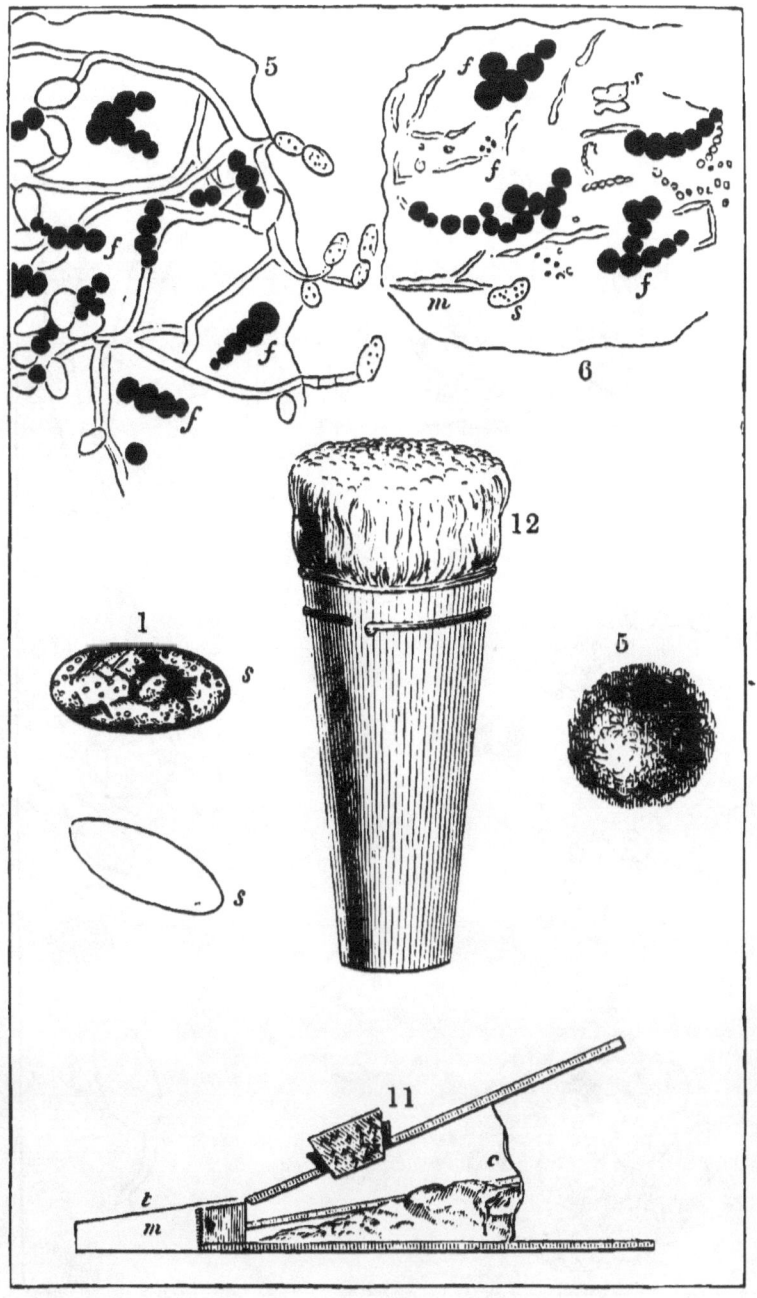

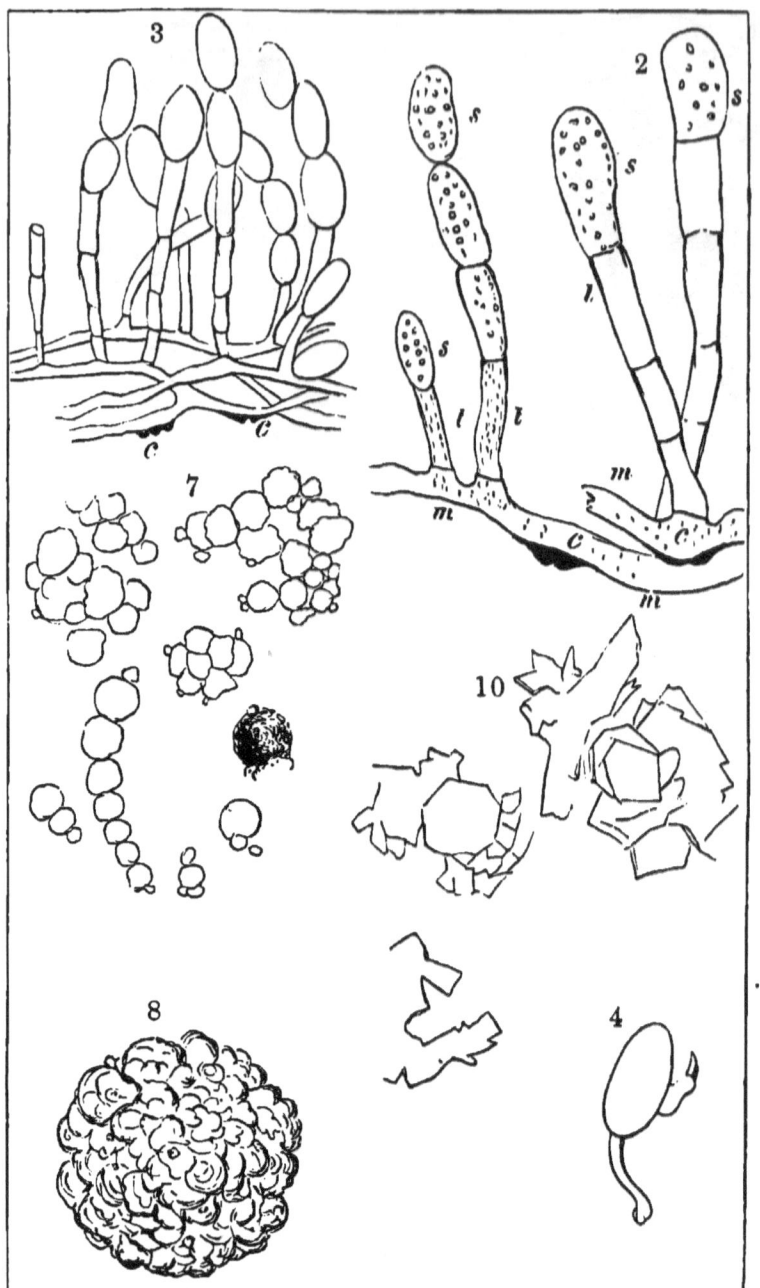

PREFACE TO THE THIRD EDITION.

When the first edition of this tract was published, in the month of May, 1856, sulphur had been employed against the vine malady by only a small number of vine-dressers, the others doubting or contesting its efficacy. To-day it is no longer so; sulphuring of diseased vines is generally adopted in practice. Thus it enters on a new phase, as any other fact does that has been subjected to rational observation, and borne the tests of criticism and experience.

To have thus quickly overcome the general resistance which new processes always encounter before being adopted in practice, it was necessary, on the one hand, that the knowledge of good methods should be diffused, in order that the operation should not be left to the hazard of empirical processes; and, on the other hand, that the reappearance and intensity of the disease should be general in the south of France, and put every where in evidence through all that region the enormous differences between the results given by vines *methodically* sulphured and those which were not.

K

The public experiments that the Agricultural Society of the Department of the Herault made in sulphuring vines in 1856 have demonstrated the exactitude of the principles I have indicated. The continuity of rains, which, by retarding the warming of the soil by the atmosphere, retarded also the general invasion of the disease, has proved how insufficient were the pretended preventive sulphurings, as well as how costly and dangerous the employment of empirical methods could become. I have therefore believed it would be useful to have printed a new edition of these instructions on the sulphuring of vines. Moreover, no other practical method has yet been discovered to supersede the use of sulphur, and to it must we still have recourse to defend our vines from the attacks of the oïdium.

This little work, having no other object than to throw light on the practice of sulphuring vines, can not be considered as a treatise on the vine-disease. As regards that, I have entered into the details and developments belonging to the subject in my "Mémoire sur la maladie de la vigne:" it can not find place in this pamphlet, where all is subordinated to the practical. If I have devoted a few pages to giving the ideas which are now entertained concerning the causes of the disease, it is simply to facilitate a comprehension of the means described, and of the method which should be followed in applying the sulphur understandingly and economically. We know, moreover, that it is useful to familiarize ourselves with new things, and to try and comprehend matters which must frequently recur, as the sulphuring of the vine must still continue to do, to the greater part of our vine-dressers.

The importance which the sale of powdered sulphur has attained in the vine-regions of the south of France has determined me to devote a whole chapter to the examination of those various powders sold under the names of flour of sulphur and ground (or triturated) sulphur, according to their origin—to the study of the conditions they must fulfill, and to the proper means for ascertaining their degrees of purity and fineness. I have also insisted, with more detail than before, on the effects which sulphur in powder produces on the vegetation of the vine and a large number of other cultivated plants.

The means which I describe in this work have all been subjected to the provings of a daily experience, from which I have each year obtained the most complete and satisfactory results. I hope the cultivator will find in the whole a guide to show him a method of operation at the same time easy and sure, and to

give him useful indications in every contingency
which may arise.   This is the end I have striven to
obtain, and I shall be happy to have succeeded.

MANUAL FOR THE SULPHURING OF DIS-EASED VINES, AND RESULTS.

THE EMPLOYMENT OF SULPHUR, AND ITS EFFECTS.

The disease of the vine was for the first time observed during the year 1845, on some stocks grown on
trellis in a hot-house at Margate, a sea-port in the
southeast of England.   It is, therefore, only thirteen
years that its appearance has been well authenticated.

It is in vain that the researches of the learned have
sought a description or a designation of that strange
malady in the texts of ancient authors, especially in
those of Theophrastus and Pliny.*   The reading of
those texts, on the contrary, shows that the authors
treated only of particular ailments to which vines

---

* Theophrastus lived in Greece about three centuries before Christ.
Pliny lived in Italy in the first century.[1]

---

[1] I see, however, that Mr. Strong, in his work on the grape, claims to have
found a description of the disease in the writings of Theophrastus, and an
account of its nature in those of Felix Fontana, who wrote about a hundred
years ago.

were subject in the climates they inhabited, or in analogous ones; for instance, the "charbon," which causes a general falling off of the blossom (described by MM. Esprit Fabre and Dunal under the name of *anthraconose*), and the scalding of the grapes by the burning heat of the August sun. Nothing authorizes us to infer that those authors wanted to describe the disease now causing so much injury in our vines. Researches in modern authors have been no less barren.

The characteristics of the vine-disease are so clearly and sharply definable, and every where present themselves with such a uniformity, that one could never mistake a description of it. Then, again, its ravages have been so great that it would in any age have fixed the attention of historians, of naturalists, and agriculturists. We should have found, then, in the authors of former centuries, positive records of a fact of such high importance, if it had presented itself. Thus we are authorized at the present time in considering the disease of the vine as a fact wholly new, whose obstinacy and extent call for the serious attention of agriculturists and naturalists.

Issuing from the hot-houses of England, the disease has little by little invaded the whole European Continent where the vine is cultivated, as well as the basin of the Mediterranean and the isles of the ocean. It was in 1851 that its appearance was positively known in the Department of the Herault, and the greater part of the French and Italian shores of the Mediterranean. Since then it has not ceased to propagate itself and work the greatest ravages.

Spain and Portugal, invaded one or two years after the eastern vineyards of Southern France and of Italy, are to-day just as badly treated.

In 1856 the product of the French vines was generally bad. In the centre and the north there was, at the same time, a notable decrease of the disease, but in the south it again came forth with disastrous power.

In 1857 the disease was general in all the south; but, energetically combated with sulphur in a great number of localities, did not prove so disastrous as in the years preceding. The malady also continued to decrease in the centre and north under influences favorable to the vegetation and fructification of the vine. Possibly, then, the period of decrease in the disease, so desirable for vine-dressers, had arrived; but the experience of 1857 again too well demonstrated the superiority of the products of the sulphured vines, both in quantity and quality, for it to be yet prudent to leave the vines to themselves. The persistence of the disease in vineyards where it has been the most efficiently combated is a fact at this day so well established, that we should continually be on our guard against it. The sure and practicable means which we possess of combating this disastrous scourge must still be called into use, and still continue to confer on vine-growing countries benefits whose value will augment with the increased care bestowed. Those means are, in their employment, based on the methodical application of sulphur in powder, to which we owe, up to the present time, the most undoubted success.

The idea of applying sulphur to the cure of the vine-disease is old; it dates almost from the appearance of the disease, for it was proposed by an English gardener of Leyton, Mr. Kyle, in 1846, and by Mr. Tucker, the first observer of the oïdium, who combined sulphur with lime; but it received then little attention.

It is in France that this application of sulphur has been really studied and propagated. Thus, in 1850, M. Gontier, the able gardener of Montrouge, near Paris, obtained from it, in his grape-houses and gardens, excellent effects, and devised the use of the bellows to administer it, after having first moistened the branches and grapes. It was this last method of

application on branches and grapes moistened be-
forehand that was practiced in some of the vineyards of the Bordelais and of the south as early as
1853, and particularly by Count Duchatel and Doctor
Turrel in 1851 and 1852, but it was under conditions as yet too embarrassing and costly for application on a large scale.

In 1853 Mr. Rose Charmeux conceived the idea of
scattering *dry* sulphur on the foliage and on the diseased fruit. This idea rendered the use of sulphur
practicable in large vineyards. It was published for
the first time, with a report on the decisive results
obtained at Thomery on 300 acres of vines, by the
Imperial Society of Horticulture of Paris, at the end
of 1853, and subsequently by M. Rendu, Inspector
General of Agriculture, in the beginning of 1854
(Report to the Minister of Agriculture). These doc-
uments, which received an immense publicity by being reproduced in all the journals, leave no uncertainty as to the value of dry-sulphuring vines. It
was only after their publication that it was employed
on the vines of the south in the year 1854.

But to obtain in every case, from the use of dry
sulphur, sure and complete results, certain determinate conditions must be complied with, which had not
by any means been sufficiently described. It was

for that reason that its employment, in the south above all, was followed now by failure and now by success, which left the question of sulphurization undecided, and gave rise to the most opposing opinions.

It needed, to decide this question, a more complete study of the disease itself and of the action the sulphur has, whether on it or on the vine, and that, moreover, precept should be united to example, and the exactitude of scientific observations demonstrated by the practice and experience of several years, in vineyards of vast extent, and under the most varied circumstances, to insure a success that none could doubt of.

This was the task I undertook to perform in the year 1855, at the time when the question of sulphurization was the most controverted, happy enough to obtain the encouragement of the Imperial and Central Society of Agriculture of Paris, and afterward that of the Society for the Encouragement of National Industry. Since then, experience has confirmed anew the results of my researches.

But, before explaining the methods of operation which I have deduced from them, it is necessary to say a few words of the march of the disease, of its essential characteristics, and of the causes to which we should attribute it. We can afterward better appreciate the action of the agent to be employed against it.

DEVELOPMENT OF THE DISEASE OF THE VINE.

The vine may be attacked by the disease at all epochs of its vegetation; it only needs a series of hot days to begin for it to appear here and there on varieties particularly accessible to its ravages, the *Carignan* and *Picardan*, for instance. If the season is cold and backward, the disease manifests itself later.

Such developments so early in the season only occur in plants already invaded during the preceding year, and belonging to varieties well known as being especially subject to the disease. Thus, as early as April, the carignan may show signs of it, but only partially, on a few buds, which will appear as if more or less powdered with flour, and which soon wither. On the other kinds it does not appear till later, ordinarily in the month of May, and even then attacks only here and there a single bud.

It spreads in proportion as the weather gets warmer, and then its ravages take a general character, no longer limited to a few isolated stalks. Thus, at the time when wheat ripens, it is seen to break out on all the plants at once. The vines then take a peculiar yellow color, and if you closely examine their leaves and fruit you will find them covered with a white dust, or efflorescence of a peculiar musty odor, easy to recognize. The invasion becomes from this moment general; it extends over immense surfaces, over entire countries, advancing a few days swifter or slower according to the heat of the climate or of the exposures.

This is at the beginning of the grain harvests, an epoch which in the climate of Montpellier varies from the 20th to the 30th of June; it is also the close of the period of blossoming. The oïdium from this time spreads every where and attacks by degrees every variety, enfeebling their vegetation and destroying their fruit during the heats of July and August. It is then only that the shoots become covered with black spots, that the leaves curl up and dry, that the grapes, at first powdered with white, become covered with brown spots, split, and dry up. Ravages of all kinds become evident to the eyes of all, and illusions and hopes founded on the vigor of the vegetation vanish before the reality.

Such is the general course of the malady; the injury it causes is ordinarily greater when a warm and dry summer succeeds to a rainy spring.

Warm, stony, shallow soils are most often those where the ravages are greatest, because the disease makes its appearance there sooner than in soils that are deeper, and therefore slower to become heated.

The dry winds and the first heats of summer ordinarily make it spread with marvelous rapidity.

Vines on trellis and other high-trained vines arc more affected than vines in souche.

Certain varieties resist better than others the attacks of the disease, and such varieties are known and noted in every vineyard at the present time.

Old vines suffer much more than young ones; their product becomes null. The old wood and branches for the most part die, and must be grafted or uprooted.

Good cultivation does not keep away the disease, but gives the vine more power to resist it.

It is the same as regards manuring; at the same time, when it heats the soil, it favors the precocious appearance of the disease, and renders it more dangerous; in such case we must be ready to combat it energetically.

CHARACTER OF THE VINE DISEASE.

Wherever the vine disease has been observed, it has been characterized by a little cryptogam or mushroom, named *oïdium Tuckeri*, after Mr. Tucker, who first observed it. The appearance of this little mushroom, similar in the first days of its development to a whitish mold, is, then, the *fundamental character-istic* of this disease. It can not be recognized on the different parts of the plant except by the presence of the little mushroom, and, on the other hand, wherever this last is established on a vine, it is diseased; so that the study of the disease itself and that of the oïdium are inseparable, and the two are confounded the one with the other.

It is only the different developments of the oïdium that give to the disease its divers aspects, according to the epoch when it is examined and the degree of intensity it has acquired.

Thus, as its seat is the epidermis (or outer skin) of all the green parts of the plant, we find the oïdium on the shoots, on the leaves, and on the grapes—on all parts that are green, in a word, and direct obser-vation shows that it is by it they are being injured. The old wood, on the contrary, and the roots, on which no particular sickness is remarked, are not at-tacked by it.

The yellow and dull aspect which the diseased vine takes in the beginning is a first symptom of the development of the oïdium; the white spots seen on shoots, leaves, and fruit signalize its presence and the stage of its very active vegetation; the characteristic

musty odor which the vine exhales is nothing else
than the oïdium itself; the gray color noticed on the
parts attacked, after a while, is owing to the state of
old age of the mushroom; the black spots which appear along the shoots, leaves, and grapes are the in-
delible traces of alterations its presence on the surface has produced, and which remain after its death
and disappearance.

The stunting of the shoots and of the grapes, the
curling up and premature fall of the leaves, the development of inter-leaves, are the consequences of
the profound and long-continued disturbance which
the oïdium has carried into the vegetation of the vine,
a disturbance whose manifestation dates from its appearance upon it.

In fine, the injuries (lésions) to be seen on the
grapes, the cracking and drying of the berries, are
the consequences of the alterations to which the ex-
terior of their tissue has been subjected: they cause
it to lose its elasticity, and to rupture when the interior parts come to grow.

THE OÏDIUM TUCKERI.

The *Oïdium Tuckeri* is a microscopic cryptogam
strongly resembling the *érysephes*, little mushroom
parasites of the most dangerous species. It is itself,

then, nothing but a parasite, which feeds at the expense of the very substance of the green parts of the different organs of the vine on which it plants itself, and thus prevents their development. Down to this time it has not been observed any where else than on the vine, and does not develop itself on other plants. The cryptogam observed on various fruit-trees, on hops, on the bind-weed, and rose-bushes, are all different from the oïdium or érysephe of the vine, although their structure may be analogous to it.

It is composed,

1. Of creeping filaments, very loose, performing the functions of roots (Fig. 2, *m, m*); they are very numerous, elongated, ramified, without cells, covering with an inextricable net-work the plant attacked. They are provided with globular protuberances, which penetrate the outer covering of the tissue, and serve as anchors to hold on by (Fig. 2, *c, c*). These last, by degrees, form about them the black spots noticed on the shoots, leaves, and fruit. This assemblage of creeping filaments is designated by the term *mycelium.*

2. Erect filaments, divided from distance to distance into cells, and club-shaped. The cells are susceptible of being transformed each into a particular kind of seed. These filaments are designated as

tigelles, or *fertile filaments,* in contradistinction to those of the mycelium, which are designated by the name of *sterile filaments.*

3. Of spores or sporules (Figs. 2 and 3, *s, s, s*); these are elipsoid corpuscles, that is to say, about the form of an egg, engendered by the cells of the tigelles, borne by them, and placed end to end at their extremities. These spores perform the function of seeds of the mushroom parasite; they germinate and reproduce it in all its parts.

Thus the oïdium is furnished with distinct organs, which fulfill the functions of roots, stalks, and seeds.

It needs a good microscope to see the oïdium well, for it is extremely small, as will be seen by the following figures, which express its dimensions.

The filaments of the mycelium are from 3 to 5 thousandths of a millimètre (a millimètre is .03937 of an inch) in thickness. Singly they are imperceptible to the naked eye, which can not see them except grouped in masses.

The tigelles have a diameter of 4 or 5 thousandths of a millimètre in the narrowest part, at the base; it is often double that at the top. Their length varies from 7 to 15 hundredths of a millimètre.

The spores are of variable sizes; in general their largest diameter is 25 thousandths of a millimètre;

it is often less. Their smallest diameter is about 10 thousandths of a millimètre.

The oïdium mould, when it covers all the different parts of a diseased vine, consists, then, of an enormous number of individuals. The fragments of the mycélium reproduce it as scions, and the spores as seeds. Each little surface covered with mould may be considered as a nursery, capable of furnishing a prodigious quantity of reproductive elements, and which the movement of the air will afterward spread abroad on all sides.

In hot and damp weather, the oïdium multiplies itself thus very quick by scion as well as seed, and can suddenly infect great extents of vine-plantation. It spreads on the surface of the green parts, and fastens there, interlacing it with a multitude of the filaments of its mycelium, on which sustain themselves, like fibres on a surface of velvet, the tigelles loaded with spores. In the early days of its appearance, the tissues on which it spreads are not impaired; but, little by little, they discolor, rot, and are destroyed. There results from this a disorganization which affects the parts over which the oïdium spreads, and especially the grapes. We may then prevent the bad effects of this parasite by attacking and destroying it as soon as we see it appear, and before it is

well established on all the vines of the vineyard;
whereas, if we wait too long, the evil is done, and no
remedy can avail.

DIFFERENT OPINIONS ON THE VINE DISEASE.

This assemblage of properties so peculiar explains
why the disease propagates itself and develops; how
it effects such ravages when circumstances favor the
vegetation of the mushroom, and do not oppose its
development. Many observers also consider this
cryptogam as the cause of the disease. Their opinion has still greater force, since it has been proved
that the disease can be cured, on parts newly attacked, by simply rubbing it off, or by destroying it in
any other mode. Thus the direct study of the disease has brought us to these conclusions:

1. That it is by the oïdium alone that we recognize it.

2. That its disappearance, wherever it has not affected the tissues, is marked by the disappearance of
the disease and of its effects.

The logical consequence of these conclusions is
that the oïdium produces the disease of the vine by
developing itself upon it, by disturbing its vegetation,
and by exhausting it after the manner of parasites.

Other opinions are brought forward; they are of

two sorts. One of them would attribute the disease to the presence of certain little insects whose stings cause the wounds on which afterward the oïdium develops itself. This system is at this day altogether abandoned, for direct observation has never been able to discover any insects whatever in numbers sufficient to produce effects like those of the disease.

The other opinion attributes the disease of the vine to a particular state of the plant itself. That diseased state of the interior that has been observed might be produced by the appearance of the oïdium, and would then be the effect, and not the cause of the disease. This explanation might be well founded if we knew to what to attribute the diseased state of the interior; but, down to the present time, nothing has occurred to justify the supposition. Vines are diseased in all countries, in all situations and exposures, in the best soils, of every species. Every kind of treatment, by manuring, pruning, cultivation, etc., have failed. We must then give up this explanation, since experience does not confirm it.

The only opinion which, down to the present time, agrees with direct observation, that which attributes the disease to the development of the oïdium at the expense of the various organs of the vine on which it plants itself, is that which, it appears to me, should be adopted.

CONDITIONS TO BE FULFILLED IN ORDER TO COMBAT THE VINE DISEASE.

In placing ourselves at this point of view, the problem of how to combat the vine disease resolves itself into that of destroying the oïdium, or its germs, in all stages of their development, and on every part of the vine where they may be found.

I have made, toward this end, for several years past, a great number of experiments of all kinds, and I have realized that it was hardly possible to destroy the disease by attacking, during the slumber of vegetation, the germs which reproduce it. The means employed for this purpose may accomplish it, and the disease will not disappear for all that.

In effect, as soon as vegetation is in movement, a cloud of reproductive oïdium germs, transported by the currents of air, light upon the green portions, take possession of them, and at the end of a few days the disease breaks out anew; the oïdium grows, fructifies, implants itself every where, destroys the fruit, and emaciates the vine.

The methods which aim at merely curing the diseased grapes are still more insufficient than the preceding, because they leave the disease full sway during the first three months of the vegetation of the vine, the very time when it is most redoubtable, and abandon altogether to its ravages the shoots and leaves. Such methods really amount to very little.

As we can only certainly know the presence of the disease by that of the oïdium, and as it fastens only on the green parts, it is upon those green parts we must attack and destroy it as soon as the parasite *begins* to make its appearance there.

The conditions to fill are therefore the following:

1. To operate on all the green surfaces of the vine in vegetation, penetrating wherever that fine dust can penetrate which forms the spores of the oïdium.

2. To renew as often as necessary the application of the destructive agent employed against the oïdium, since the means of reproduction it possesses are incessantly at work, and it can develop itself anew as soon as the green surfaces of the vine cease to be protected from its attacks.

3. To apply the remedy before the oïdium has been able to impair the tissues of the different parts of the bud—above all, when it is young. This last condition is the most indispensable, because, if we fail to destroy the parasite until it has more or less affected the parts, we shall obtain but a partial result at best—the evil is already done.

These three conditions should be fulfilled by means that are sure, practical, not too costly, and which do not interfere at all with the divers operations of cultivation.

PROPERTIES OF SULPHUR.

The object is attained in an admirable manner by the flour of sulphur. It possesses, in fact, all the properties necessary to constitute it the curative agent "par excellence." On the one hand, it destroys the oïdium whenever coming in contact with it; and, on the other, its form, being that of a very fine dust, enables it to envelop by a simple aspersion the entire plant in vegetation, and its volatility in the temperature daily produced by the heats of summer on the earth and the green surfaces exposed to the sun insures its action on the mischievous germs. It has, besides, the property, as remarkable as unlooked-for, of stimulating the vegetation of the vine, thus communicating to it vigor to conquer the attacks of the parasite.

Sulphate of lime, soda, potash, which destroy very well the oïdium when they can be brought in contact with it, are not at all volatile like sulphur. They have not at all, as it has, the property of penetrating in the form of vapor all those places left un touched by it when in the form of dust, and of continually renewing their curative action by every day vaporizing a little. They have, too, the serious inconvenience, if they pass into the wine with the grapes, of imparting to it a bad taste not always to be got rid of.

The mixtures of sulphur and of dust, such as pulverized earths, plaster, etc., have the inconvenience of neutralizing, more or less, the action, of the sulphur, so that by using such we risk obtaining but incomplete results. We may, besides, injure the quality of the wine if the mixtures in question are capable of forming soluble combinations in it.

ACTION OF SULPHUR ON THE OÏDIUM.

By direct observation under the microscope, we are able to see that the grains of flour of sulphur cause the oïdium to perish when they enter in contact with it.

One condition seems always necessary, which is, that the temperature should be above 20° of Centigrade (68° of Fahrenheit) when the contact takes place. Now this condition is always filled during days of sunshine, from the time the buds begin to put forth in April and May. Later, during the days of summer, the temperature almost always passes this limit, even in the shade. As the oïdium does not propagate itself nor develop rapidly until the temperature reaches 25° to 35° of Centigrade (77° to 95° of Fahrenheit), such a heat insures the action of the sulphur against every increase of the disease. When the temperature is too low, the sulphur does not sensibly act; and if blown off by wind or washed off by rain, it must be applied anew. The action of the sulphur on the oïdium is quick enough, but it does not become apparent until after a few days.

When the sun strikes the diseased parts which have been covered with sulphur, the action is much more energetic and rapid : it becomes apparent from the second day, and often sooner. This results from the warmth of the sun's rays.

A sulphuring well applied, which reaches the entire surface of the vine, will therefore destroy the oïdium; but, as the vine grows continually, and the grapes enlarge daily, as the wind and the rain carry off all the while some of the sulphur from the surfaces where it was deposited, they soon become bare again and are exposed to new attacks ; then the oïdium again appears and attacks the vine, as it did at first. This occurs ordinarily in summer time, after an interval of twenty to twenty-five days,* or some-

* When the ground is very damp beneath the surface, and the lat-

times a longer one. Then, again, it may happen that the oïdium will not make a second appearance in serious form, especially if the weather continues for a long while very dry and very hot, and the sulphur rests a good while on the ground and on the various parts of the plant.

The high temperature produced by the vertical rays of the sun in summer vaporizes the sulphur in a perceptible manner. It gives out then a very lively odor, which all must remark who have employed it on their vines. It is that portion which falls on the ground without reaching the vines which feels more particularly the effect of the heat and passes into vapor, because the sun's action heats the soil much more than it does the foliage. It results from this that the sulphur which was spilled and seemed lost, produces, on the contrary, the happiest and most continued effects, by passing daily into vapor under influence of the daily sunshine. Its molecules thus penetrate

ter is dried, baked, and covered with weeds, and great heat succeeds to heavy, drying winds, the oïdium will reappear more quickly. In such circumstances I have seen, in 1856, in the month of July, its invasions renewed after only ten days of interval. Such a combination of circumstances is not frequent; at the same time, it will occur in wet seasons. In such cases the sulphurings must be brought closer together, and renewed as often as the invasions are repeated.

L

numberless points on the foliage and fruit that might not otherwise be reached.

I am sure that it is not at all to the sulphurous acid, nor to the sulphuric acid which is found in small quantities in flour of sulphur, that the action of the last upon the oïdium is due. The same effects may be produced after having washed it: they may also be obtained with sulphur-rolls well pulverized.

[Here follows a very closely detailed examination into the characteristics of the two forms of sulphur, namely, the ground and the sublimated, or flour of sulphur. The chapter devoted to this is omitted, with the observation that flour of sulphur is the best to use in America, as the difference per cent. in price is not here so great as it is in France, and its greater cheapness there is the only advantage it possesses over the other. The flour is very light and fine of texture, and of a beautiful yellow color. Ground sulphur is much the lighter in color, and the finer (and, of course, the better) it is, the lighter is the hue.]

SULPHURING OF DISEASED VINES.

The sulphuring of vines is an operation which consists in spreading over their foliage and fruits sulphur in fine powder.

Flour of sulphur obtained by sublimation is ordinarily the form of it most suitable for the purpose.

Three conditions are necessary to insure a good result.

1. The application must be made *as soon as the oidium begins to appear on the vine.* Thus the parasite will be prevented from obtaining too strong a foothold on fruit and foliage to impair their vegetation and disease their tissues.

2. The sulphuring must be renewed as often as the oidium renews its attack, and as soon as it reappears. Thus we continue to operate on the surface of the plant, and to prevent the bad effects which would not otherwise fail to follow a new invasion.

3. The application should be thorough, and reach every infected part. It will not do, therefore, merely to sulphur the diseased fruit; the shoots, leaves, and all the fruit—in a word, every green part, must be dusted with sulphur. When we find a single bud or a stalk to be diseased, we may be sure every other bud carries on its surface the germs of the disease; to destroy these germs, they must be reached with sulphur-dust.

The fundamental principle is this : *scatter the sulphur on every green part upon the first appearance of the symptoms of the disease, and renew the ap-*

plication each time it reappears. It is especially at the commencement of vegetation that we must keep the vines free from attack. At that epoch the least delay is dangerous, for the buds then attacked are so young and feeble they have no power of resistance; they emaciate, are stunted, and very quickly lost. Moreover, the earlier the disease appears the more virulent it generally is. The Carignans and Picardans, varieties which are usually the first to be attacked, are examples of this.

A precautionary application may be made before any sign of disease appears. It can do no harm, but there is no rule for prescribing it. Such applications have no other objection than that, if made after vegetation is well advanced, they cost something. In the earlier periods of the development of the buds, that is to say, about the middle of May, the cost would be insignificant. But preventive applications should not be relied on too long, nor lull the vigilance with which we should watch every portion of the vineyard.

Such a surveillance it is easy to organize by dividing the field according to the varieties it contains, and keeping a memorandum of the applications each division has received.

Proprietors who have their vines sulphured should

make sure it is properly done. It is an operation that needs the eye of the master.

Sulphur may be applied at all hours of the day when it does not rain; it is indifferent whether the surfaces where it lodges are wet or dry—the action is the same; provided the temperature is not below 20° Centigrade (68° Fahrenheit), it will destroy the oïdium wherever it touches it.

At the same time, the best conditions for employing sulphur, and for its quick and lively action, are a dry and hot day, a brilliant sun, a light wind to aid its dispersion without disturbing that operation, and dry surfaces to receive the powder.

In my practice five applications have sufficed to combat the most malignant cases of the disease, where the attacks begun on the "Carignans" the 2d of May, and were repeated down to as late as September. On some young Aramons not invaded till July, only one sulphuring proved sufficient.

Between these two extremes, I have found that in the greater number of cases two or three sulphurings arrest very well the effects of the disease; three have been enough in the greater portion of my vines, and, notably, in 1854, when the malignity was remarkable.

Following the principles I have set forth, I employed powdered sulphur in 1854, 1855, 1856, and

1857, on great extents of vine plantation, situated on the most varied soils, and growing all the varieties cultivated in the south of France; and I obtained so complete a success that there was no exception to it even when the disease had attained its greatest virulence.

The empirical methods that have been applied to sulphurization will never give such a collective result. They have led into error a crowd of operators by designating in advance a fixed epoch for employing the sulphur and the number of sulphurings necessary. This has been because the disease varies so much in the periods of its appearance and successive reappearances, according to soil, exposure, variety, and culture, that no general epoch can be fixed on of practical application to all varieties and all soils, etc.; we may hit right with some, we will miss with others. Moreover, these methods have produced a mixture of good and bad results which served to cast a cloud of uncertainty upon the efficacy of sulphur in the first years of its introduction, and retarded its general adoption.

MANNER OF APPLYING SULPHUR TO VINES. — INSTRU-
MENTS MOST SUITABLE FOR THAT PURPOSE.

We can, if we choose, fling sulphur upon vines
without any instrument, or we can dust or dredge
it over the foliage from a thin muslin bag; a sieve
may also be made to serve; the effects will be good
if·the operation in any of these ways be but carefully done. But the waste is great in such case, and a
strong wind arrests the work entirely. It is therefore important to adopt a good instrument.

The quantity of sulphur necessary might be almost indefinitely reduced if the sulphur were itself
divided to infinitude, and if we could spread it with
perfect regularity and uniformity, for great masses
of it are not needed to produce the desired effect on
the oïdium: it suffices that the dust of this substance,
no matter how small its grains may be, should penetrate wherever the oïdium or its germs may be.
With powder perfectly divided, and a suitable instrument, a great economy of sulphur may be realized, then.

The instrument should satisfy the following requi-
sites:

1. Throw the sulphur-dust far enough, and scatter
it uniformly, so as not to fall in lumps.

2. Be able to augment or diminish at will the issue of dust.

3. Not be inconvenient to the person who uses it.

4. Facilitate the work.

5. Be easy to handle, and capable of being used either by men or women, or by children of twelve years.

6. Be strong, and not require repairing.

7. Be of sufficiently low price.

Down to the present time the instruments in use do not meet all these requirements; but, if they have not yet attained the perfection of implements that have been subjected to the tests of a long experience, they can, nevertheless, do good service.

Those instruments are the following:

1. Bellows of different kinds.

The box-bellows, at first used, consisted of an ordinary bellows, on the nozzle of which was fixed a tin receptacle, which was traversed by the blast, and into which the sulphur was put.

These bellows projected the sulphur with force, and sent it well among the foliage, but a great deal of it went out in lumps; they held but a small charge, and needed frequent refitting; they were heavy, and hard to handle, because the load was carried at the extremity; they often got out of order from the weight of the receptacle, which works apart the joints, and fastenings, and ends, by wrenching off the nozzle.

The Vergnes bellows has replaced the above with advantage.

It is a common bellows without a valve, and whose whole interior serves as a receptacle for the sulphur (Fig. 11). At the base of the nozzle (f) is a sieve with large meshes (s), and made of coarse tinned iron wire (m): the sulphur is thus prevented from going out in lumps, and the tin preserves the iron from too rapid corrosion.

The air comes in and goes out by the same way, by the nozzle, which should have at the base a sufficient diameter to be slightly conical.

A two-inch hole is cut in the upper board of the bellows to receive the sulphur, and to this hole a stopper of wood is fitted (b). A pound of sulphur is a proper charge, and this will dose fifty vigorous vines as they are in July, when the shoots cross one another and entirely cover the ground.* Care must be taken not to overload the bellows, because then it can not play easily, and the leather soon bursts near the hinge.

* A South of France vine in this stage of its growth would take twice as much sulphur as one of ours would.—F.

L 2

The advantages of the bellows I have just described are these:

It is cheaper than those made with a tin receptacle between the body and the nozzle.

It is more manageable, because all the weight is near the hand.

It can carry a larger charge, and therefore needs less labor.

It flings and scatters the sulphur, which, kept constantly in motion by the entrance and exit of the air through the same opening, is better divided, and makes fewer lumps.

It is necessary that the leather of these bellows be of excellent quality and very strong; inferior leather is soon corroded and covered with holes.* They can, besides, be very well and cheaply mended when holes appear, by pasting pieces of leather over the holes with a strong mucilage, which, in fact, makes them stronger than before.

* Here I foresee a difficulty. With the best tanning material in the world, we have the worst leather in the world. Owing to a want of the critical faculty, as well as of economical foresight on the part of the consumers, and a want of conscience on the part of the tanners, we lose every year, from wearing bad shoe-leather, a sum sufficient to pay the interest on our debt.

I think this is the second time in the course of this volume I have provided for that interest; and, now we have chosen Grant for President. I hope he will attend to this matter.—F.

These bellows, nevertheless, have always the inconvenience of letting out some of the sulphur in lumps, and of breaking in holes.

2. The perforated box.

This is simply a round tin box, slightly conical, furnished at its larger extremity with a double bottom pierced with holes for the sulphur to sift through and pass out. The smaller end has a cover fitted to it, to be opened when the instrument is charged. This is the simplest and cheapest of all the instruments, and has the great advantage of never getting out of order; but it has numerous difficulties: it is fatiguing and troublesome to the workman, works slowly, badly distributes the dust, and letting it out in lumps, causing great waste. Of all instruments in use, this one renders sulphuring the most dear and least expeditious.

An improvement on the last is the box with a tuft (Fig. 12). The same in all other respects, it differs only in having a tuft of wool attached to the surface of the perforated end, the shreds of which are four inches long. This tuft becomes filled with sulphur, and gives it off when shaken, receiving continued supplies from within the box. It lets out no lumps, and diminishes considerably the expenditure of sulphur; it avoids, thus, the two great disadvantages of the simple perforated box, but does not remedy others. Besides, it can not be worked except with a very small quantity of sulphur, for otherwise it would gorge, and let none out; it therefore needs frequent filling, which wastes time.

Down to this time I have used, notwithstanding its inconveniences, the Vergnes bellows, and it is with this instrument the results have been obtained which will be found farther on.

Doubtless better results as to economy might be obtained with a more perfect implement. But, until sufficient practice shall have enabled me to record my testimony in favor of such a one, I limit myself to pointing out in what direction we ought to search after such improvements, and in designating such instruments as have been more particularly brought before the attention of the public.

OF THE EPOCH FOR APPLYING SULPHUR.

If the opinions of even the most eminent cultivators is so divided as regards the efficacy of sulphur against the disease of the vine,* it is because that, to obtain of that agent all its good effects, it must be applied at a determinate moment, which varies according to climate and variety. It is from not hav-

* There is no longer any such division of opinion in France.

ing operated properly, or not having sufficiently renewed the operations, that they have not succeeded.

The moment to seize upon is, I repeat, that when
the first symptoms of the disease appear. If we
wait too long, and the oïdium gets a strong hold on
the vine, we shall never completely cure its attacks.
With certain varieties the grapes will be destroyed
(Piquepouls and Terrets); with others they will be
profoundly injured—they will become brown, crack
open at the season. of maturity, etc. People will
then say that sulphur has no power over the oïdium,
that they do not believe in it, that it is better to do
nothing, etc. One accustomed to see diseased vines
learns very soon to seize the favorable moment; and
here is the way to recognize it, and then to treat it,
as it is manifested on our principal varieties in the
south of France.

[Although it may seem at first not worth while to
read the details which follow respecting the different
methods rendered necessary by the many different
varieties of vines grown in the south of France, yet,
upon reflection, and especially by reading those details, the American vine-dresser will recognize the
value of every thing they contain. They will show
how one variety is afflicted in one way, and another
in another—how wide is the range of the manifesta-

tions of the oïdium, and how various—and prepare him to deduce principles of diagnosis and treatment for each of our own native kinds, as each in turn shall become infected in its own peculiar way.]

Carignans.—Of all varieties of the vine, the Carignan is that which the disease attacks in preference; in a great number of cases it takes the infection first, and then communicates it to the other varieties in a disastrous manner. In this respect it may be considered the *seed-bearer* to the mushroom parasite. It is it that spreads and disseminates the disease at the epochs when it is most particularly dangerous. It merits, therefore, very particular attention.

When a Carignan is attacked by the disease, we see, here and there, shoots affected by the oïdium as early as in April, only a few days after their putting out. These buds are covered, wholly or in part, by a whitish or gray dust of a characteristic musty odor. At this epoch we must search closely to find these, for they are still very young and relatively few, and do not strike the eye if it is not practiced in watching for them. Thus a vine of Carignan will appear in good condition to the eye of a not very attentive observer, which is, at the same time, completely infected with the malady and already badly injured.

At this epoch (April and the first days of May), Carignans thus invaded have not the yellow and sickly hue which the oïdium brings when it comes later in the season. As soon as we perceive the young buds to be affected, we must apply sulphur, and unsparingly pluck off such as are entirely covered with the oïdium. Strictly speaking, they may be cured, but they are already so impaired they will always remain stunted, and can give but an insignificant product.*

From this epoch (May), if the sun shines on the sulphured vines during several hours, or if the sulphur remains forty-eight hours on the vine without being washed off by rain, the effect is produced. It will not become evident till several days afterward. We will then find that the affected shoots have lost the peculiar musty odor which accompanies the oïdium. On examining closely the young leaves, we will see that the white spots have become gray, that they are no longer accompanied with dust, and have lost their odor. The disease is then suppressed for

---

* In 1854 I cured shoots of the Carignan so completely attacked as to be covered as early as the 12th of May with a gray powder : they were hardly four inches long. Their grapes ripened perfectly healthy, but the shoots, although cured, remained stunted. It sufficed to keep them constantly sulphured during the first month, and afterward renew the application every fortnight.

about three or four weeks. At the end of that time the shoots have grown. If it is perceived that the vine takes a sickly yellow hue, that the young leaves of their extremities are covered on their under sides and about their edges with white spots, that they are slightly crisp, we may be sure the disease will not be long in making a new irruption. We must then again make a general application of the remedy on the shoots, leaves, and grapes. It will act like the other if the weather is warm enough, and if a rain does not immediately carry it off. Its effect, if the sulphur remains several days on the vines, will continue at least three weeks.

The surveillance must continue from this time onward. A third, fourth, and even fifth application must be made, if necessary.

The Carignan, while it is one of the vines which take the disease the most severely, is, at the same time, one which sulphur preserves the best. Under its influence it gives magnificent products.

We always find a few Carignans scattered among mixed plantations. They there propagate the disease in a disastrous manner. To prevent this, it is enough to sulphur them at first by themselves once or twice, if need be, at the beginning of vegetation; afterward they are treated with the rest of the vineyard. This method is not expensive, and succeeds very well.

In 1855 I preserved the worst attacked of my Carignans with five sulphurings, applied at the following epochs: the 2d of May, 19th May, 5th June, 9th July, and 14th August. In the deeper soils, the disease being less precocious, three applications, made at the end of May, June, and July respectively, were sufficient.

The Aramons.—The Aramons are attacked later than the Carignans. It is hardly before the second fortnight in May that we find appearances of the oïdium on their shoots. These are, as yet, few in number, and scattered here and there. Before the end of May they must be pulled off and sacrificed. The vine will be infallibly attacked later, either in June or July.

The signs which warn us the invasion has begun are a decided yellowness of the leaves, accompanied with little white spots around the indentations of the young leaves at the end of the shoot; a slight crisping of those leaves; efflorescence on the berries of grapes or on their stems. This is the time to make a general and thorough application of sulphur; its effect will be sure. In about eight days after the vine will take a beautiful green color; all the efflorescence will have disappeared. This arrests the disease for twenty or thirty days, according to the weather. After that interval, signs of a second attack will be indicated by the shoots becoming yellow; by white spots on the lower side of the young leaves; by the whitening of a grape here and there. Then give a second sulphuring.

If this is too long delayed, the yellowness increases; the leaves crisp; all along the shoot what we call *inter-leaves* put out: these are new leaf-shoots, slender and curled. This last sign is characteristic of the disease, like the musty odor I have named. Finally, the white spots on the grapes cover them entirely and become gray. Then the fruit is injured, and will bear traces of damage even after being cured: it will be likely to split when it ripens. Sulphur, in this condition of things, may re-establish the vine, but will not preserve it in its full vigor, as it would if applied a few days sooner; and to accomplish even thus much, it may be necessary to renew the application from week to week.

The sulphur should be blown from a bellows, and be made to reach all green parts, but, at the same time, be more particularly directed toward the grapes. If the second sulphuring has been properly done, and toward the end of July, no other will be needed.

In general, two sulphurings have been enough to preserve Aramons strongly invaded ; in a few cases only I have had to give three, and on young vines of this variety, not attacked till late in July, only one has sufficed.   I sulphured my Aramons in 1854 and 1855 : in the first year, on the 9th of June, 14th of July, and 1st of August ;* and in the second year twice, namely, from the 9th to the 11th of June, and from the 10th to the 12th of July.

It is noteworthy that the grapes of the Aramon resist strongly the disease, while the wood is easily affected by it.   We also notice Aramons that preserve and ripen some fruit, although their young wood is seriously attacked.   The vine is none the less diseased for all that, and will, in general, produce much less fruit the following year.

Alicantes, Aspirans, Mourastels, Œillades, Bruns Fourcats, Clairettes, etc.—We observe and treat the greater part of these red varieties as we do the Aramon.   With them the disease has the same phases— the yellowness, crispness, inter-leaves, and invasion of the grape—but, in general, with less distinctness of manifestation.

* In 1856 my Aramons were sulphured, mostly, from the 24th to the 30th of June, and from the 15th to the 25th of July.   Only a very few of the vines needed three applications.   The oïdium came later than it did the preceding years, on account of the continuous rains.

The *Alicantes* and Mourastels resist the disease remarkably well.

The Muscat.—This vine has in its form and in its wood a great analogy with the Aramon, but it does not resist the oïdium well, and its fruit is affected much sooner than that of the other.

The *Piquepouls* should be especially watched. As soon as the disease appears on them, so abrupt is its attack, the crop is ruined in a few hours if measures are not taken. The attack aims principally at the grapes, and is much less felt on the shoots and leaves, which is the reverse of what happens with the Aramon. After the beginning of June or last of May, whenever we see the verdure of the shoots grow pale, and a few spots appear on the young leaves, sulphur must be applied; it will arrest the disease for three or four weeks. It must be afterward renewed according as it may appear necessary. I have succeeded in saving Piquepouls strongly attacked by means of three sulphurings properly administered in May, June, and July. Under the influence of sulphur, this variety gives very beautiful products.

The *Terrets* take the disease in an anomalous manner. Sometimes they resist very well, and give the best crops; sometimes they give, on the contrary, the worst results, because the disease takes the form of "*rougeau.*" We seldom see any of their shoots diseased until the epoch of general invasion comes, which is about the time of blossoming (the 25th of June), or a little after, from the 1st to the 5th of July; sometimes it comes sooner, in the first fortnight of June; sometimes, again, it does not come at all, and then the vine is naturally cured. This last case is rare, however, and it is best not to count on it.

The invasion announces itself by a slight yellowness of the leaves; these, at their extremities, show white spots, chiefly at the indentations. On the grapes, before as well as after the blossoming, oïdium dust appears. If the grapes have not blossomed, the injury may already be considerable.

Rougeau of Terrets.—On the appearance of the earliest symptoms, sulphur must be applied without delay. A few days of delay are very prejudicial, for the disease of the vine takes suddenly, with Terrets, a new and terrible form: the vegetation of the plants ceases, the leaf turns red, dries, and falls; the fruit dries or becomes atrophied; often it continues slowly to become covered with a gray dust; the berries then detach themselves one by one, or cease to grow any more: hardly a trace of fruit remains.

It is this particular form of the disease that we

generally designate at this day, in the Department of L'Herault, under the name of *rougeau*. This does not at all resemble the peculiar affection that the vine-cultivators of the north and east called *rougeot* before the disease of the vine was known. Messrs. Dunal and Esprit Fabre, who first called attention to the rougeau in January, 1854, took great care to distinguish it from the other, but they did not attribute to it the origin which I give it, for they made of it a disease entirely distinct from that produced by the oïdium. Since 1853 I have declared the rougeau of Terrets to be a particular form of the vine disease. It is produced by it, because we never see the rougeau appear on our Terrets before the invasion of the oïdium—it always follows that, and is its consequence. Besides, when sulphur is applied to the Terrets at the moment of the oïdium's appearance, the disease is cured, and the rougeau does not appear.

A good application of sulphur at the beginning of the disease averts it for about twenty-five days. After that interval, if we have not yet reached the time when the grapes begin to color (the 15th or 20th of August), we may again perceive, on the young leaves at the end of the shoots, and on the grapes, new beginnings of the disease; often even the signs of rougeau itself. We must be prompt to sulphur again, which will carry us safely through to vintage.

When the application of sulphur is too long delayed, and the *rougeau* actually comes, we must not hesitate to have recourse to the remedy anew, unless things have gone so far that all is lost. In 1854, from the 17th to the 19th of July, I sulphured gray Terrets strongly attacked by rougeau—the crop was already half destroyed. At the end of ten days vegetation resumed its course, the rougeau was arrested, and all the grapes which had not become atrophied were preserved. The vines in question yielded half a crop. Those in the same field which were not sulphured perished entirely.

The rougeau works its ravages principally at the epoch of our greatest heats, from the 15th of July to the 15th of August. Of all forms of the disease, it is the most disastrous and most rapidly destructive.*

In general, two sulphurings, well done, and at the proper moment, combat it effectively on Terrets, although it often assumes a disastrous intensity. This

---

* Possibly this form of oïdium is the one for whose maw our dear little Nortons are destined. Certainly something very like the rougeau came to my Norton vines last year. We hear, too, of the Delawares losing their leaves. We shall see.—F.

is because the oïdium begins its work upon that variety quite late.

In 1855 and 1856 two sulphurings proved sufficient to preserve my gray Terrets. The first was given between the 4th and 7th of July, the second between the 1st and 3d of August.

VINES SULPHURED THE PRECEDING YEAR.

Every year since 1855 I have made comparative observations of vines sulphured in 1854, 1855, 1856, and 1857, and those which were not. In both, the oïdium made its annual reappearance at the same epoch, and with the same intensity.* In 1854 I sulphured several vines from the 7th to the 17th of August. This was late enough, certainly, for the operation to have a preservative effect; yet in the following year, as early as May, I found the oïdium on shoots of those vines. What sufficed to meet the disease and render it harmless for one season was not sufficient to prevent its recurring early in the next; so that the use of sulphur from year to year is not a preventive against the disease; nor is it any more so when resorted to at the beginning of vegeta-

* In 1856 Carignan vines sulphured and preserved in 1855 were attacked after the end of April, like others not sulphured at all. In 1857 they were attacked in May.

tion. The proof is, we are obliged to keep renewing it, if we would obtain good results. This simple statement is enough to demonstrate how slight are the grounds on which all pretended preventive methods of employing sulphur rest. Those who have advocated them have never succeeded in preventing the coming of the oïdium. They combat its effects as they are combated where rational means are used, but with this difference, that they employ, without motive, much useless material. Sulphur is simply a destructive agent *par excellence* of the oïdium, inasmuch as this last dies, as we have seen above, when brought in contact with it. Apart from the impulsion it gives to the vegetation of the vine, the action of the sulphur against the oïdium is then curative; it has no other character.

The vigor of sulphured vines produces generally in them, the following year, a more abundant production of fruit than that of such as had suffered by the disease. It is now a well-recognized fact that the shoots of such as I have last named present, in the season of putting forth, a quantity of grapes much less than that of vines in a normal state.

M

PRECEPTS TO FOLLOW IN APPLYING SULPHUR TO
DISEASED VINES.

It is well to observe the following precepts in applying sulphur to diseased vines:

1. The vines attacked by oïdium should be cultivated with special care; we should leave no weeds about them; the earth should be kept always loose. Every thing that enfeebles vegetation favors the action of the disease; for instance, bad pruning, plowing too seldom or doing it badly, the washing away of the earth from slopes, etc. The invasion of the parasite mushroom troubles profoundly the vegetation of the plants. They must be reanimated by cultivation, while at the same time the parasite is destroyed by sulphur. In this way the most complete results will be obtained.

If a diseased vine is manured, it must be cultivated and sulphured with particular care.

2. It is better to apply the sulphur too early than too late.

3. Sulphurings at the moment of the blossoming have appeared to me the most efficacious; they appeared, besides, to exercise a salutary action on that phase of vegetation. I thought I observed, in 1854 and 1855, that vines which received the sulphur at

that epoch had "knotted" their berries better than the others.* It destroys the oïdium at the moment when it is most capable of injuring the grapes, and the effect is therefore all the more valuable. There is no vine-dresser who has not seen, in certain years, grapes of the Terret disappear in a few days from having taken the oïdium at the moment of blossoming. The presence of sulphur prevents that disaster.

4. Every sulphuring should be carefully made, and reach all parts of the plant—shoots, leaves, and fruit. To spare the flour of sulphur is bad economy. The dust should be flung on either by walking all round the plant, or by doing first one side and then the other. The work is well done when, taking a bunch or a leaf, and holding it between the eye and the sun, numerous grains of fine dust can be seen upon it. Always bear in mind that sulphur destroys oïdium only when brought in contact with it.

5. When a vine has been sulphured, it is proper to wait some days, at least, before plowing. What of the powder falls to the ground should be allowed to

---

* I observed the same thing in 1856. It was confirmed the same year by the experiments of M. Cazalis-Allut; and this important fact appears to me now beyond doubt, and deserves place among those the most interesting to viticulture.

volatilize in the sun, and rise and condense on the shaded portions of the vine; it will thus penetrate daily the sheltered places where a simple aspersion of the dust would not have carried it. To turn it under with the plow would defeat this process.

6. If a rain comes and washes off the sulphur the very day it is applied, there is no risk in waiting a few days before renewing it. Notwithstanding the rain, the effects of the first application are considerable, provided the temperature has attained 68° to 85° Fahrenheit. From the time the vine is well in leaf—the month of July, for instance—strong rains, even, can not remove the sulphur, and even in May and June they derange less than was at first feared.

7. The conditions most favorable to the action of sulphur are hot and dry weather and a clear sun. Nevertheless, sulphur may be applied in all weathers, and nothing should stop it, when the need is urgent, unless it be rain.

If it is needed without delay, wind should be no objection. I have sulphured, in June, vines poorly leaved, during high winds, and succeeded well. In such cases it is only needful to use a little more material than in a calm.

8. The effect of a sulphuring can not be judged until after about ten days. It is requisite, in fact, to

give the vegetation time to regain its normal progress and to develop anew. A rain falling a few days
after the sulphuring renders the good effect of it
much more conspicuous. The whole vine takes a remarkable greenness and brilliancy : the leaves seem
varnished.*

9. The sulphur is no preventive; the need of renewing it so often proves that. If we would proceed with that economy of which large agricultural
operations are worthy, it is proper to await the symptoms before resorting to the remedy.

10. After August 10th, in the climate of Montpellier, the effect of sulphuring on the varieties of red
grapes is hardly appreciable in preserving the fruit.

11. When the grapes begin to color without being
attacked by oïdium, they are out of its reach. If
they are already affected before beginning to turn,
they will continue to suffer. What precedes has explained why sulphuring performed toward the end of
July, and done just when it should be, and as it should
be [I use ten English words to translate " à-propos"],
carries the fruit safely through into the vintager's

---

* A gentleman, writing of the oïdium in South Carolina, notices
such a varnished appearance of the leaves as *preceding* the development of the parasite. The two facts are worthy of being compared,
and may possibly bear on the question of the origin of the pest.

basket. This is a fact confirmed by experience ever since the disease became known. The epoch of coloring in the Department of L'Herault (according to years, varieties, and exposures) is from the 5th to the 25th of August.

OF THE QUANTITY OF SULPHUR NECESSARY FOR THE TREATMENT OF DISEASED VINES.

Cost of Sulphuring an Acre in May.

12 pounds of flour of sulphur at 13½ centimes......................$0 32½

Wages of a woman at 1 franc per day of 8 hours............... 16

Total...............$0 48½

Cost of Sulphuring an Acre in June (15th to 20th).

40 pounds of flour of sulphur at 13½ centimes.....................$1 08

2 days' labor of a woman, 8 hours effective at 1 franc.......... 40

Total...............$1 48

Cost of Sulphuring an Acre in July.

Aramons, the most vigorous.

48 pounds of flour of sulphur at 13½ centimes* $1 30

Labor of a woman 3 days at 1 franc.................................... 60

Total...............$1 90

In most kinds of vines, sulphuring in July costs no more than in June, and $1 48 may be taken as the

---

* Five centimes equal one of our cents. The cost of sulphur in Cincinnati, by the wholesale, I found to be 7½ cents, nearly thrice the cost in France. Men's wages in that country being usually 50 cents, and in ours $1 50, it appears we should multiply by three the above estimates of M. Mares, and by four if men do the work.

cost per acre from June onward. For two applications, then, the cost will be $2 96, and for three, $4 44. The quantity of sulphur used will be from 80 to 120 pounds per acre. This last figure is the *maximum*, and is rarely reached; in most vineyards the *minimum* is seldom exceeded. Better sulphur than I have used would cost more, but less of it be needed. The dryer it is, the farther the same quantity will go, and the less will be the labor required.

Practically there is no more simple operation than sulphuring vines, even when they are in their fullest luxuriance. Where the vines are trained to stakes, as in Bordeaux, Champagne, and the Bordelais, far less material and labor are needed. Comparing the results obtained with the expenditure, no operation can be more advantageous; the vines are preserved on the soil, and their products saved from the worst scourge that has ever attacked them.

CONCERNING THE VEGETATION OF SULPHURED VINES.

The effects of sulphur on the vegetation does not begin to be appreciable until the end of spring, or in summer, about eight days after the application. Then the branches are seen to recover their beautiful green color and to vegetate with new vigor. At each application the same effect is manifested in a marked manner; the vine also maintains a vigor so well sustained, provided it be well cultivated, that its fruit ripens much more equally and much earlier. These facts are at this day beyond doubt in all places where the proper method has been carefully followed.

I have before alluded to the favorable influence of sulphur on the blossoming. I again observed it in 1856, and the same year the observations of many cultivators confirmed my own, which I had published the year before. This fact, so important, is, moreover, not isolated, nor does it apply solely to the blossoming and fructification of the vine. I ascertained, in 1866, that sulphur favors the fructification of a great number of fruit-trees, particularly plum, quince, pear, and apple trees, and exerts on the vegetation of a large number of cultivated plants a powerful influence. From this point of view we may easily arrive at the conclusion that vines should be sulphured when in blossom, whether the oïdium is present or not.

A more even ripening, and probably, also, a special action of the sulphur on the coloring matter of the grapes, makes the wine from sulphured grapes have a higher color, so that in the departments of the south such wines maintained over the others an incontestable superiority in 1855 and 1856, and it was the same in 1857.

Sulphured vines have every where preserved their leaves with remarkable persistence, such as was only observed in well-manured vines before the disease came. In 1855 and 1856 they kept their verdure up to the frosts of December, looking like so many green islets among the others, despoiled of their leaves since the month of October. Their wood is healthy, beautiful, and very much developed. Their products in grapes have been those of good years. Their wood being very vigorous, the shoots present the following year a show of fruit more abundant than vines that did not take the disease.

The effects of sulphur on diseased vines is really marvelous when it is applied "*à propos*," and often enough to prevent any ravage of the oïdium. The same vine, divided in two equal parts, has given me, according to the virulence of the disease, two or four times more fruit on the sulphured part than on the other; the difference of product in grapes being still greater when we operate on Carignans, Piquepouls, etc.

This remarkable vegetation of sulphured vines brings us naturally to put the question, Is the sulphur a manure, or at least a stimulant for the vine?

M 2

What I have observed down to this time, and particularly in 1856, after having put the question in
the first edition of this work, leads me to answer af-
firmatively. At the same time, it will not do to conclude that the use of·sulphur dispenses with that of
manure. I have noticed that its good effects tend
to diminish when the soil is neglected as to manuring, while in soils well manured and well cultivated
its effects on vegetation sustained themselves during many succeeding years in the most remarkable
manner.

I feel sure that sulphur augments considerably
the vegetation and fructification of the vine independently of its state of disease. It will be a valuable agent to increase the fertility of vineyards and
render them more regular, but upon condition that
manure be concurrently employed, otherwise its action will yearly grow less, and end by becoming insignificant.

This same consideration ought to reassure those
who think the stimulating action of sulphur may
soon exhaust their vines ; for that stimulating action
can not exert itself except in so much as it is favor-
ed by the richness of the soil, and will not, to any
considerable extent, increase the fruitfulness of bad-
ly-kept fields. In those well kept up, sulphur acts

like good cultivation and manuring, which, while developing the productive force of the vine, are far from exhausting it.

In any case, nothing is more worthy of interest than the study of questions arising out of the use of sulphur to stimulate vegetation. It is a wholly new field, in which the student of vegetable physiology and agriculture may find numerous subjects of observation.

REVIEW OF THE CHAPTERS WHICH RELATE TO THE EMPLOYMENT OF SULPHUR AND ITS ACTION ON THE VINE.

In this review we see that the action of sulphur on the vine is exerted in two ways quite distinct.

In the first place, sulphur destroys the little parasite mushroom (oïdium Tuckeri) whose development on the branches and fruit constitutes the disease.

In the second place, sulphur acts by exciting vegetation and.favoring fructification.

Thus two distinct properties reciprocally complete each other when the disease is to be combated.

The sulphurings should therefore be so timed and regulated as that the one should work to the advantage of the other.

Thus we should always sulphur at the period of blossoming, in order that the fructification should operate more completely, and this without regard to the disease. If, before or after this is done, the oïdium makes its attack, it will be sufficient, independently of the sulphuring at blossom-time, to give another each time the first symptoms of disease show themselves. I have described them minutely, and, besides, every one knows them too well at the present day to make any mistake.

I have shown by numerous examples that in the most obstinate cases it is rare that four sulphurings are necessary, not counting that given at blossom-time; that oftener it suffices to add one or two only to this last, in order to obtain the best results. We have also seen how the surveillance of a large vineyard, of varieties the most diverse, is made easy, by keeping a memorandum of the fields in cultivation, and of the sulphurings given them; for the study of the oïdium proves that its reappearances are almost always separated by an interval of twenty to thirty days, according to the intensity of the disease, the variety, soil, cultivation, and temperature.

Thanks to this system, I have indicated how the application should be made to all the varieties of most importance in the South of France; how we may be always in time, always sure to succeed. We realize the greatest possible economy, and do not, all at a time and needlessly, employ an excessive force of laborers at seasons when they are so scarce that often the most urgent labors must go undone.

For those who would free themselves from the trouble of watching over their vines, and care little about the additional expense — who prefer a rule ready made—there is another manner of procedure equally sure : *it is to apply sulphur to their vines every twenty days, beginning at the moment when the shoots have attained the length of two inches, and continuing till the grapes begin to color.* In the climate of Montpellier these two epochs are comprised between the 1st of May and the 10th of August, or thereabout. In that interval of a hundred days there will be five or six sulphurings to make, whose effects will be assured, as well for the purpose of destroying the oïdium as for that of favoring the vegetation and fructification of the vine. By this system, which is based on the interval which separates ordinarily the reappearances of the oïdium, the average cost of material and labor will be double that which I have given as the highest estimate.

OBJECTIONS AGAINST THE USE OF SULPHUR.

The principal objections made against the employment of sulphur are the following :

1st OBJECTION.—*The good effects are doubtful.*

If the good effects are doubted by any, it is because they have not made the application under the proper conditions; either their vines had been too long invaded, or the sulphuring was not soon enough renewed and at the opportune moment, or it was incompletely done, and directed against the diseased grapes only instead of all the fruit and entire foliage. It is probable that later, when the employment of sulphur shall have become more general, and we understand better its management, no one will seriously dispute its value.

In all the course of this work I have demonstrated by direct observation, or by practice and the production of facts, how ill founded is this objection. I will add to these a few considerations to answer the arguments of those who tell us they every year see diseased vines cure themselves spontaneously without any sulphur, and that if those which are sulphured recover, it is not the sulphur that cures them, but Nature.

There are vines, it is true, which rid themselves spontaneously of disease; nevertheless, they are never entirely delivered from it, and for that reason are not comparable for strength and beauty to the same vines treated with sulphur. We know them as soon as we look at them. We always see numerous traces of the oïdium on their grapes, but principally on their shoots. If they pass for being. spontaneously cured, it is not so much because we see no oïdium appear upon them as because they give a better crop than in preceding years. It would be more proper to recognize in their condition a marked amelioration than a cure. The number of vines really cured spontaneously is small.

But, in any case, spontaneous cure and sulphur cure are two things which do not conflict at all, especially when that which passes for spontaneously cured, like that which is cured by sulphur, is subject to be again attacked by the disease. Finally, comparative experiments made on the same vine divided in two parts, of which the half which was sulphured was perfectly preserved, while the other half, left to itself, was completely ravaged, answer all objections. Such comparative experiments were often repeated in 1854, 1855, and 1856.

2d OBJECTION.—*The use of sulphur is expensive.*

I have replied in advance to this objection, and proved that such expense hardly amounted to a fourth or a seventh of the ordinary current expenses of a vineyard, including interest, taxes, etc. It is, nevertheless, no trifling expense, but it is very largely compensated down to the present time by the increased price obtained for the wine.

3d OBJECTION.—*Sulphur used on the grapes imparts a bad taste, which enters into the wine.*

This is no better founded than the other objections. It is true, wine from grapes recently sulphured has a very marked sulphurous flavor, which lasts a good while, but when nothing else has been mixed with the sulphur, there is nothing to fear from this; ordinarily it passes away without leaving any trace after the drawings off which remove the wine from the coarse lees. The taste, besides, is slight when the sulphuring has been done in a rational manner, without covering the grapes with a useless quantity of dust, and that, too, so late in the season that no good could come of it, however applied. I have had in my cellar many large casks full of wine from sulphured grapes; it has always kept well and brought a good price. Nor have I ever learned that the case was different in the cellars of others. To my view, the feeble quantity of gas which may develop from the sulphur in such wines is a preservative agent, and imparts a peculiar stability.

But, however this may be, there is a sure means to avoid the contracting of any such taste. It consists in making the last sulphuring not later than from the 20th of July to the 15th of August, and in using a bellows instead of flinging on the powder from a box, and at a period so late that there can be no need for it in any case.

But, even after the sulphur taste has been contracted, it can be removed easily enough by once or twice drawing off. Usually the first drawing off, which, by removing the coarse lees, disposes of the greater part of the sulphur, is found sufficient.

M. Barral has pointed out a way of more promptly effecting the same object: it consists in drawing off the wine into a cask in which sulphur has been burned in the ordinary method of fumigation. The sulphureted hydrogen which gives the bad taste decomposes in contact with the sulphurous acid introduced by the fumigation, and the wine is quickly relieved of the presence of the former. Wine destined for the still should be carefully rid of the taste in question, as otherwise it will enter into the brandy.

We know that great quantities of sulphur are used by wine-merchants to fumigate their casks. There is no more reason to fear injury to the quality of the wine in the one case than in the other.*

* But a few years since, wine was shipped from Cette to Holland quite new, on the lees, and strongly sulphured. This wine, after the treatment, appeared to have lost color. In that state it made the sea

In 1855 and 1856 wines from sulphured grapes generally possessed a superiority over others which caused them to be sought after. Their color was livelier and their maturity more equal. It was the same in 1857. These are advantages which decide the question altogether in favor of sulphur.

When the sulphur used on the vines has been the triturated article from raw material of inferior quality, or mixed with other matters capable of forming sulphurets, it may happen that a disagreeable taste will be produced, quite distinct from that which comes from sulphureted hydrogen. This will be due to the presence in the wine of a small quantity of soluble sulphurets: it is very tenacious. At the same time, if the casks are strongly fumigated with burning sulphur and drawn off several times, it ought to disappear, because sulphurous acid decomposes soluble sulphurets.

When vines are treated with sulphur that is free from all mixture, like the flour or triturated rolls, there is no danger of such accidents.

4th OBJECTION.—*Sufficient sulphur can not be obtained to cure, every year, all the diseased vines.*

voyage, and on its arrival was allowed to repose. It was drawn off several times and then clarified. It became excellent, and was remarked for its freshness and delicacy.

Admitting that the vine disease will continue for a long time and with its present virulence, we shall certainly need a good deal of sulphur. But there is hardly a limit to its production; millions of quintals may be obtained, if needed. The rise in price of the flour did not extend to crude sulphur, so large is the supply. It was the suddenness of the demand that made flour of sulphur temporarily dear. Let the extent of the prospective demand be known, and the increased means of manufacturing it will bring down the price to a reasonable point,* and medicine enough be found to heal all the sick.

* Within six months from writing the above, this prediction was fully realized.

CHAPTER XVII.

MONTPELLIER AND HENRI MARÈS.

AND now we will go to Montpellier and make the acquaintance of M. Marès himself. Montpellier was once a capital of Languedoc, Toulouse being another. As long ago as the times of the Crusades, it was a republic, free and flourishing, and "the hope of the free." It was ruined, however, by long wars, waged on each side to correct the theology and save the souls of those on the other side.

In a fine old mansion near the beautiful park, or rather point of view called the Peron, I found M. Marès. He met me as he should, and I spent an hour with him explaining what my errand was, and endeavoring to interest him in it. On parting, he promised that, on the day but one following, he would call and go with me over some of his vines, excusing himself for not doing so the next day, as it was a fête.

To properly occupy the fête-day, I took the train for Cette, the neighboring sea-port, a place of large wine commerce, and famous for its "imitations," as they call the Madeira, Sherry, Port, etc., they make there. These are not, however, like those of Chaptal and Gall, imitations of the miracle of Cana, but are based on sound, strong, natural wines, which are flavored and named in deference to British and American tastes.

The train stopped for a while at the village of Frontignan, famous for the Muscat of that name. The Muscat requires a soil open, warm, and, at the same time, rather strong. The grapes must hang on the vines a certain time after maturity, and on any but a dry soil would rot during the heavy rains of the last of October, which I have said were common in Languedoc. To give sweet and high-flavored wine, the vines need to be at least twenty years old, which is a pity, for we could make Muscats grow in many parts of America, and the wine, so luscious, and, at the same time, so delicate, would suit the tastes of our people, and help them learn to love wines in general. It would be an admirable sugar-teat to wean teetotalers.

In different parts of L'Herault as many as 5000 acres are planted with this variety, but the average yield is very small, in some neighborhoods not exceeding sixty gallons to the acre. It is grown in

souche, or, if trained to trellis, yields an inferior product. The spaces are only three feet.

I walked about the streets of Cette, and went
through some of its vast warehouses. Nobody would
tell me what were the ingredients put in to make
Port, Sherry, Madeira, etc. They manage to do without cellars by calling by that name their ground-floor
store-houses, and sprinkling their pavements in hot
weather.

During the evening, after returning from Cette, I
invited to my room the old, well-decorated officer of
the First Empire who kept the hotel. He said the
vine product had greatly increased of late years, in
consequence of the rise in prices, which still kept
rising, notwithstanding the increase. But the wine
was not what it once was; the resort to manuring
had not only lowered the quality, but so affected its
keeping properties, that whereas formerly good wines
would keep fifteen years, now they last only five. The
better kinds, St. George's, for instance, used to bring
only about 18 cents a gallon; the price of the staple
red wines of L'Herault now ranges from 10 to 25
cents. And this is the story all over the French
vine districts — increasing consumption, advancing
prices, extended plantations; manuring, increased
yield, and inferior quality.

Next day M. Marès called for me, and conducted me over one of his vineyards lying at the edge of the city. The wine-house belonging to it was the same in which Chaptal, at the time a resident of Montpellier, carried on his experiments in falsifying wines. His example has found few followers in his own province, however, and the only wonder is that, with real wine flowing like a sea around him, and selling for 5 cents a gallon, he should have thought it worth while to employ his great abilities in making it from water.

The vines I found to be furnished with from six to eight shoots or canes to each souche, each cane cut back to two eyes and what they call a sub-eye. The very old vines had more canes than any others. To prune, they use a two-handed shears, and make a square cut, instead of one which leaves the tip in shape of a whistle. These shears are said to save three fourths of the labor, and, since the cost of pruning is so large a proportion of the whole expense of working vines in souche, it is said that, but for their invention, wines of the south could hardly have kept their place in the market.

It is difficult to see how vines in souche could be pruned any longer than they are, if the fruit is to be kept off the ground. As it is, however, all the clusters being collected about the crotch of the souche, and balancing each other from opposite sides, they are very well sustained. It is likely many of our American varieties would prove too long-jointed for such training, or bear their fruit so far out on the branch that, in cutting back to two eyes only, we should cut away all the fruit-buds. There are vines in France which have this habit, and for such M. Guyot's system might be a good one. Some of the plants I saw had the valuable peculiarity of holding their canes, and the shoots from them, quite erect, while others drooped and trailed.

The field had received its first plowing. Good cultivation, M. Marès said, required three workings of the ground in a season, and sometimes four. The first should be done between the first of January and the last of March; the second, between the 15th of May and the 1st of June; and the third, some time before the 24th of June, or be deferred till the first fortnight of July, though generally that is rendered impossible by the vines spreading so as to completely cover the ground. Formerly twice working was considered enough. Hand-labor is thought the best, and doubtless it is, but the plow is more generally used because it is more economical. It is not unusual to plow both ways, the furrows crossing each other.

The first operation is often so performed as to throw the earth away from the feet of the vines, leaving them standing in little trenches seven inches deep. As this is done at any time from the first of January to the last of March, it would be a destructive practice in our country of cold winters, as one of my neighbors across the Ohio learned at his own expense, when a very skillful vine-dresser undertook to do in Kentucky what he had been taught in France.

Naturally one of my first inquiries was for the reason why low-souche training was not seen in other parts of France. The reply was that, except in the warm climate of the south, fruit would not ripen unless spread out on trellis or stakes. The Folle Blanche, however, which thrives in the comparatively cold and damp region of the Bordelais and the brandy country of the Charente, is an exception to this rule.

"These vines," said M. Marès, as we passed into another plantation, "in their third season from cuttings, which was last year, yielded 100 hectolitres to the hectare" (about 1000 gallons to the acre). Older plants in the same field had given double that quantity.

Seeing bits of rags sticking out of the ground here and there, I asked if they had been brought

N

there as a manure. They had, and rags are held in very high esteem. Oil-cake, too, is considered particularly valuable.

In remarking on the space allowed between the plants, M. Marès said that experience had shown that nothing was gained by setting them any closer, an acre of vines standing within three feet of each other yielding no more wine than if they were five feet apart.

Early on the following day, which was the 15th of April, M. Marès called by appointment to drive me out to his large estate, ten miles from Montpellier. On my remarking that the morning was sharp and frosty, he told me there had been a severe frost both the last night and the one before, and he was rather anxious to learn how far his vines had suffered; but he was apparently less preoccupied during the drive out than might be expected of a man with 250 acres of vines and two sharp frosts on his mind.

An hour and a half or less brought us within sight of a large chateau, with a wall about it, amid plantations of olives, mulberries, and vines, otherwise uninclosed. Halting the carriage a good way short of the house, we got out and entered a field bordering the road, and the proprietor began his inspection. It did not by any means please him, and the result was that he became convinced he had lost half his prospective crop on all vines sufficiently advanced to be hurt.

" We are ravaged, M. Henri !" exclaimed, in a bluff voice, a bluff, wholesome-looking man, with a gun in his hand, who came to meet us. It was the overseer, who was, at the same time, the "*frere-du-lait*" of M. Marès. In old countries, where they try to remember and not forget, the relation of foster-brother (brother of the milk) is rather a near one, and I am sure the overseer performed his duty with more fidelity, and more pleasure too, for considering himself a member of the family. He had been through the fields since daylight, and made his report as above— " We are ravaged." " I think so," was the reply of the other, as he continued on, tramping over the dusty furrows, continually stooping to examine the buds as he went, like one who is his own chief overseer. And I must tramp after him and listen to all he said, for seldom could I obtain admission to a lecture such as he was giving. We talked as continuously as we walked, his familiarity with every detail, and his scientific knowledge of his subject, more than ever convincing me I had fallen into the hands of the right man. I could easily see, also, that the fields were clean of weeds, every vine-souche in good health, with none missing, the alleys well kept, and all things thing fully up with the season.

Three hours of such walking and talking brought us to the gate of the chateau and the hour for breakfast. Vines in low places had suffered the most, and in this respect it was the same with the mulberry-trees. Recently-plowed fields, other things equal, had decidedly been worse hurt than those early plowed. Late pruning was also shown to be a protection. The dangerous term, as regards frosts, is, it seems, from the 1st of April to the 10th of May (which last date is precisely our limit in the Ohio Valley), but now and then the limit is exceeded in Languedoc as well as Ohio. For the above reason, no plowing is done, if it can be helped, between those dates.

Pruning is by many deferred to the latest safe moment, and by some is done even as late as the 1st of April—"in the flow of the sap," as it is called—which is said to guarantee the wines through the whole of that month. But this practice is objected to as having a stunting effect. A good many vines had been burst (in their last year's wood) by the late severe winter, the mercury having fallen as low as 17° *above* zero of Fahrenheit. One of our American winters would astonish them, I think.

We were ready for our breakfast when it came in —for a "breakfast with forks," as they call it. While dispatching the excellent and substantial one spread before us by an old domestic, we tried several kinds of the finer wines of the country, such as Muscat, Picardin sweet and dry, etc. Five francs a bottle was the price of one of the Muscats, grown during the hottest season in memory. The price of the average quality of this kind of wine, when old enough for drinking, is but thirty cents. It is sad to know this, and then to think of the ten and fifteen fold prices wrung from our toiling thousands of money-makers, bankers, jobbers, and stock-jobbers for inferior liquids by heartless hotel and restaurant keepers.

In reply to an inquiry from my entertainer if I really thought the vine could be made to succeed in America, I told him yes, of course; it was native to every inch of our soil, found in all our forests from North to South, whereas, I went on to add, neither France nor any other part of Europe could claim it as indigenous, and for this last gave Humboldt as my authority. To my surprise, M. Marès refused to admit my fact and Humboldt's, which I and H. had affirmed a thousand times without meeting with contradiction. He assured me that wild vines still existed in many parts of Languedoc; that their fruit was even used for wine-making by the poor people; that, so wild was their nature, when attempts were made to cultivate and prune them closely in the usual mode, they would blossom, but never bear; and, finally, for illustration rather than proof, referred to the twisted column often seen in European architecture as copied from the wild vine in its gigantic forest growth.

I ventured to boast a little of the vigorous growth of American vines, but here again M. Marès took me down, and I fell heavily. "Come this way," he said, "and I will introduce you to some of your compatriots." And he showed me to his nursery, or experimental vineyard rather, where were growing a great number of vines assembled from all parts of Europe, some from England even, with three from America, the Catawba, Isabella, and York's Madeira, or Canby. The last stood in the same rows with scores of others of the same age, namely, three years, and had been treated in all respects in the same manner as the others. They were as large as a large thumb, which was certainly very well for them, but the others were as large as a wrist.

I gave up.

"They resist disease remarkably well," observed M. Marès.

He showed me his collection of plows, chief among which was the old classic article, one handled, and all of wood except a sharp-pointed, triangular plate of iron, which ran flat to the ground. I attacked the plows, thinking I could safely attack a two or three thousand year old invention. But M. Marès, while admitting the superiority of modern plows, of which several of very good construction were in use in Languedoc, insisted with many reasons that for certain kinds of work, and in certain soils, the "araire," as it was called, remained still a desirable implement. It recalled our Western shovel-plow, which is likewise of wood except the shovel plate, which does but shallow work, and stirs the soil without turning it over, and is, in fact, a step backward toward the old araire, which last, however, goes down as deep as eight inches in the first or winter plowing.

After the plows I was shown the bellows used for sulphuring. There were several kinds, most of them being known in America; but it is worthy of remark that M. Marès, after fourteen or fifteen years' experience with all of them, preferred the simplest and cheapest, namely, that described in his pamphlet as the Vergnes bellows.

A serious defect in all bellows I have seen since

my return to America is that their sieves are double, and are too fine. They should be single, and the meshes as coarse as consists with economy of sulphur.

We next looked in at the wine-cellars, or rather houses. These were of a grandeur befitting an average vintage of three hundred thousand gallons. The several apartments were furnished with fifty-five casks, called "futs," of capacity varying from seven to ten thousand gallons, besides many enormous vats, each wide and deep enough to drown very comfortably dozens of naked villains who might attempt to bathe in it. But the thing is never needed to be done in Languedoc, where grapes get so dead ripe as to need no crushing even, but are commonly flung into the vat without stemming, and with no other crushing than what they must needs get in handling and transporting. For the same reason, seven days only is the term of the fermentation.

From 250 acres in vines M. Marès often gathers as much as 15,000 hectolitres, or 375,000 gallons. The variety called Aramon, in the proper soil, gives 1200 gallons to the acre as an average.

In Languedoc, as elsewhere on the Mediterranean shores, and to some extent also in Burgundy, it is an immemorial custom to sprinkle on the grapes in the vat before fermentation begins a layer of plaster of Paris. In the south they generally make the layer thick enough to amount to two pounds or more for every hundred gallons. This is to prevent the formation of acetic acid, and is claimed to have other advantages as well. Many, however, are beginning to oppose the practice. Its opponents, while admitting that plaster helps the wine to keep and deepens its color, insist that it hurts the quality, which accusation its advocates in their turn stoutly deny.

During fermentation the vats are usually covered with loose boards. A plan for improving inferior musts has been invented by M. Marès, which is as follows: The best grapes, usually ripening ten days in advance of the others, are gathered and fermented first, and as their quantity is usually less than that of coarser qualities, the vats are only partially filled, so as that each one shall have its even share. On the eighth day the clear wine is drawn off, leaving in the bottom of the vat, however, not only the pomace, but also one fifth of the liquid. The inferior vintage is then flung in upon this residuum, and, fermenting there, takes up a portion of the virtues of the superior one, working an amelioration that sometimes doubles the market value of the wine.

This is, in fact, Gall's method, with only such difference as must always exist between real wine and sugared water.

Though seven days is the usual term allotted for wine to remain upon the skins, it varies somewhat in certain parts of Languedoc. In the Gard, for instance, it is often extended to fifteen days, and again, where it is not intended to make wine of commerce, or when the "Rose" wines of the Rhone hills is sought to be imitated, a much shorter time is allowed, ranging from thirty hours up to three days. To hasten fermentation, a part of the must is sometimes boiled and then returned to the vat.

Common wine is only once racked off, which is done at the end of winter, and in cold weather, if practicable, but in some places it is not drawn off at all; but Muscat is racked four times within the first three months from vintage.

The wines of the South of France vary in alcoholic strength from seven up to sixteen per cent., the strongest being Muscat and Picardin, both of them white. Common red wines of commerce range between ten and twelve per cent. before the dealers take them in hand to manipulate for market. These brandy heavily all they send away, which notoriously hurts their delicacy. So does the long fermentation on the skins, they say, which is also done to make the wine keep. What is retained for home consumption is, however, left in the vat only long enough to get a good color. As such wines will not bear transportation, they must be drunk at home to be truly appreciated.

From the wine-houses we looked in on the silkworm nursery, and afterward walked out to where they were building drains of stone. In Languedoc it is customary to drain heavy soils, but the custom is by no means so uniform as it is in Médoc, Burgundy, and Alsace, where more valuable products are obtained. So important do I now esteem it to drain all clay soils destined to be planted in vines, that to my mind the mere absence of drainage in the vineyards of the Ohio Valley would suffice to account for their poor success, were no other cause to be found. Deep digging, trenching with mattock and spade two feet deep, or deeper, have a certain effect for a few years, but in time the adhesive soil gets packed again, and the expensive preparation is as good as lost. And I can not but think that good tile or stone drainage would dispense with the costly trenching or deep plowing we have been used to think essential, so far as that a good turning up with the common subsoil plow would be found sufficient.

We did not get back to Montpellier till some time

after dark.   At my hotel I bade good-by to the gentleman from whom I had learned so much, and the
next day was on my way to Avignon.

A traveling companion who lived near Avignon
informed me it was the custom in his neighborhood
to heap up about the feet of the olive-trees, every
autumn, little mounds of earth fifteen inches high,
to protect against frost.   He also said they did the
same with their vines, which otherwise could not
support even the mild winters of the south.   Avignon, it will be noted, is in the great South of France
vine-country, and the vines alluded to were grown in
souche.   This was what I had not learned before,
and came from a stranger, yet I am inclined to believe it, though I don't think it is true of the vines
about Montpellier, or those immediately on the coast.
At Avignon the rail route toward Lyons enters the
Valley of the Rhone, and, for a long way up that
valley, the same method of training and the same
vines are found as those I have been writing of, and
as this continues all the way to Valence, which is in
the latitude of Bordeaux, I can readily believe in the
necessity of covering up in winter.

CHAPTER XVIII.

WOULD LOW SOUCHE VINES DO WELL IN AMERICA?

AS regards California, this is no longer an open question. In certain parts of Texas, too, they have a wild vine which takes the low souche form of itself, by help of winter killing, which regularly cuts it down to a few eyes close to the old stalk. Those of us who would try the experiment should begin with varieties whose joints are short, whose canes are stiff, or, what is better, erect in their growth, and whose fruit-buds are found close to the old stock, or souche. If we have none which combine these qualifications with the other essentials of a good plant, means can probably be found for educating such as we have into the requisite habit of growth.

In view of the possibility that we may not be able at once to lay our hand on precisely the right kind to begin with, I have imported the French varieties already mentioned.

Of course they must be covered in winter, except the Folle-blanche, which may be hardy enough to do without it in some of our Southern States; but cov-

ering little ten-inch stumps would be a very trifling
matter in comparison with the laying down and bury-
ing of high souches, which is even now the practice
in our colder grape regions, and has always been done
in some parts of Hungary.

For ripening grapes on vines trained in souche, the
requisite amount of heat during the growing season
may be estimated from the following data, obtained
after much search in a corner of the Imperial Library
at Paris, where were deposited a few volumes of re-
ports on the statistics of some of the departments.
Perhaps in a future edition I shall be able to furnish
something more satisfactory than the range of the
thermometer in but two of the departments of the
great vine-region of South France, one of them covering only two years, and the other only one. The
mean temperature of the Department of the Gard
during each month of the growing and ripening season, for the years 1838 and 1839 respectively, and
that of the Department of the Bouches-du-Rhone,
where Marseilles is situated, during the corresponding months of the year 1821, were as follows:

	April.	May.	June.	July.	Aug.	Sept.	Oct.
Gard, 1838......	53.24	63.86	72.05	70.70	74.48	67.73	61.70
" 1839......	55.58	64.86	76.17	78.26	76.04	68.38	60.20
Bouches-du-Rhone, 1821	50.	52.70	62.6	65.3	69.20	57.20	56.30

With the indication thus given of the temperature of the warmest portion of the South of France, every one can compare that of his own particular section, and judge if its climate is warm enough to ripen grapes on vines trained in low souche. I will, however, give the mean temperature of one point in the Ohio Valley and one in the Lake Erie region:

	April.	May.	June.	July.	Aug.	Sept.	Oct.
Cincinnati	54.10	63.60	71.40	76.50	74.20	66.	53.20
Kelly's Island ...		57.53	68.65	74.01	72.41	64.94	53.16

The degree of heat needed to ripen the Folle-blanche in souche is less than what the other varieties seem to require. It alone, so far as I could learn, flourishes in that form as far north as Bordeaux, and even farther north in the Department of the Charente, where, as we have seen, it yields the wine of which Cognac brandy is made. The mean temperature of the Charente, derived from observations made during a series of four years, at the hours of seven A.M. and two and *eleven* P.M. each day, is as follows:

April.	May.	June.	July.	August.	September.	October.
47.90	53.81	58.19	60.87	59.28	57.68	47.98

Whether we compare with the South of France, then, or the far more temperate region of the Bordelais and the Charente, it will appear that throughout the greater part of the United States we shall have a sufficiency of solar heat for ripening grapes on vines in souche.

It has been objected to this kind of training that it can not succeed except in an extremely dry climate. But vines in souche seem to do as well in the Valley of the Rhone, where the mean rain-fall for the year is 36 inches (the same as in L'Herault), and for the summer months 9½ inches, as in Gironde and Charente, where the yearly quantity is but 24 inches, or in parts of California, where the summer mean is less than 2 inches. Or, if we regard the dryness or dampness of the air merely, and not the rain-fall, we find such vines supporting as well the aridity of the South of France, where there are but 77 rainy days in the year, as the moisture of the Valley of the Gironde, where there are 141, or of the Valley of the Charente, where there are 150 of them in a year.

A good deal has been published in America on the climatology of the grape, in which the quantity of the annual as well as of the summer rain-fall is treated as being very important. It seems to me that the true inquiry should be how dry, or how moist, are the soil and the air during the growing and ripening process? Languedoc, with 36 inches of rain-fall, is a dry region, chiefly because those 36 inches fall in 77 days, while the Charente is a moist one, with only 24 inches of water, chiefly because it continues to come down during 150 drizzly days. The fact is well known that, as compared with the climates of France, Germany, and other wine-countries of Europe, our own is remarkably dry.

If it be said that to be safe from the oïdium we must take refuge in regions where the rain-fall measures only just so many inches in such and such months — that we must abandon our Ohio Valley and fly to the lake shore, I will ask, Why is it that oïdium in all its forms is far more pestilential in the South of France, which is very much the dryest portion of the kingdom, than in any other part of it? Or, if it be replied that this results from training in souche, I rejoin that they have there both stakes and trellis, and that it is precisely the vines trained upon stakes and trellis that are afflicted the worst.

There exist in America three conditions which render training in souche more suitable for us at the present time than the other modes. These are:

1st. Dear labor.

2d. Cheap land.

3d. Immediate need for much cheap wine.

In view of the first of these, I would observe that . the cost of creating an acre of vineyard, planted in

the fashionable varieties, and furnished with wire
trellis, exclusive of the price of the land, fencing,
and a good many other smaller items, has of late run
up to the large sum of $600, $700, and $800 in Con-
gress money, which, being reduced to the currency
of the Bible and the Constitution, means, we will say,
$500. In many places this is doubled, and in others
much more than doubled, by the speculative prices
paid for the ground. And the yearly expenditure
for attendance, merely, is estimated, I see, as high,
sometimes, as $150, or, I should say, $110 of true
money, in which we will hereafter continue to make
our estimates, if you please.

Now if I should assume the first cost and subse-
quent maintenance of a vineyard in souche to be
only one third of the cost where the methods in
vogue are followed, I should probably come as near
to exactitude as estimates usually bring us. But I
prefer each one should estimate for himself. The
same preparation of the soil is needed as with other
vines. Cuttings, and not roots, should be used, and
cuttings for an acre should not cost over $10—we
used to sell them for $2 50 the thousand. Only
winter-pruning is needed, and no pinching, rubbing
off, willow-tying, straw-tying, or leaf-pruning; and as
regards plowing and cultivating, the same labor good
farmers bestow on their corn-fields will suffice.

In view of the other two conditions, I recommend the selection of rich, warm, and, at the same time, easily tilled soils, such as are more readily found on plains than on hills. A good sandy loam, planted in souche with Concords or Ives's Seedlings, well cultivated and well sulphured, ought to bring an average annual crop of one thousand gallons to the acre, which should sell, while new, not for $1 50, $2, $3, and $4 per gallon, but for 25 cents at the very outside.

We can, and we will, grow wine cheaper than the Europeans, and for the same reason that we can grow wheat cheaper than they, namely, that we have cheaper land and more of it. In raising grapes on our present system, however, we abandon the only vantage-ground we possess, and enter into competition with them in a field where they are stronger than we.

As long as our wines, no matter how inferior, sell for a dollar a gallon, expensive vineyards, with their costly culture, may do very well, but how long will this last? More than one authority entitled to respect have lately estimated the extent of our present plantation at from one to two millions of acres. I think this an enormously large estimate, but don't doubt we shall, before very long, have a million acres in bearing, chiefly of coarse varieties and yielding large crops. We shall have a glut of wine, as we have twice had of petroleum, and a like fall in prices. This is what I want to see, and what Mr. Longworth hoped for and labored for, till the blight came upon our vines. Yet, when such a glut shall come, the sufferers will be those whose vineyards cost to make them, land included, a thousand dollars the acre, and to maintain them a hundred dollars yearly, rather than those who shall follow the more economical plan.

Fine qualities, where they happen to be on fine soils, will bear the highest possible cultivation. Of such there is no danger we shall ever see an over-production. I know of but one man in America who has turned away from rich hill-side land, and gone and planted his vines on a meagre soil, with the purpose of obtaining a choice rather than an abundant return.

For cheap culture and large crops we should go to the plains. For a small but valuable product we may resort to expensive garden-culture, if we can find an exceptional soil, and such will more easily be discovered on hills than elsewhere. To adopt expensive methods on strong hill-lands, only to grow coarse and cheap wines, is a great mistake, into which our instructors from foreign countries have led us.

CHAPTER XIX.

HOW THEY PLANT VINES IN SOUCHE.

FOR the details I am able to give on this subject, I am largely indebted to the writings of M. Marès, as well as to his verbal instructions.

Unless the soil be very light indeed, draining is essential. If this be done, a good subsoil plowing is all the preparation needed, unless it be manuring. If it is not done, the ground must be broken up to the depth of two feet. In Languedoc they go down to the depth of three feet sometimes, but usually from 16 to 24 inches is all. To break up an acre two feet deep requires the labor of two yoke of oxen and two drivers for three days, and of one yoke and one driver for the same time, the work being done at two operations, the heavier plow following the lighter.

The usual distance of five feet between the plants is, strange to say, extended to nearly six feet where the land is poor. It is thought that a wider space than five feet on rich ground induces a too great development of wood and leaves, at the expense of the ripeness and excellence of the fruit, a thing not to be feared where the ground is poor. As obviously there would be a better access for the plow if the space were enlarged, it has been attempted to obtain that advantage by planting in rows 7 feet 3 inches ، apart, the plants being distant from each other in the rows only 3 feet 7 inches; but this has not succeeded. The same variety, growing on the same soil, has been found to give twenty-five per cent. more wine where planted 5 × 5, than where planted 7.3 × 3.7.

They use both rooted plants and cuttings. Where the soil is not too dry for them to take root easily, which is, however, most commonly the case, cuttings are preferred. In Médoc, where such a degree of dryness is not to be feared, they decidedly prefer cuttings. For the larger portion of our country I feel sure cuttings are best. This is especially true since nursery-men have learned the forcing process. But cuttings should be carefully chosen, and as carefully prepared and set out. A cutting should never be taken from a barren vine nor from a barren cane, but from a fruit-branch of the preceding year that has borne fruit, and from a healthy plant of full age. In Languedoc they bring vines into bearing as early from cuttings as from roots. Where the intention is to plant in the early part of winter, they soak the lower eight inches of the cuttings for a week before planting them; but this can be dispensed with if recent rains have moistened the soil, or if the planting is to be done in the spring. Before setting them out, the bark of the cuttings is scraped in places here and there between the eyes for eight inches of the lower portion, the knife penetrating to the inner bark. Under very favorable circumstances, the failures of cuttings thus prepared, and which were planted as soon as cut from the vine, have been known to amount to only two per cent.

In Médoc, if cuttings are to be planted before their buds put out, they prepare them by setting them in a trench inclined at an angle of forty-five degrees, and cover the lower half with earth; or, if they are not to be planted so early, a ditch two feet deep is dug, on the bottom of which there is flung eight inches of loose earth; on this eight inches of cuttings are laid horizontally, separated from each other by more loose dirt sprinkled in, and finally covered with about eight inches of earth, which completes the filling up of the trench.

Another plan, mentioned in M. Du Breuil's work, is to bury them in a trench in a perpendicular position, points downward and butts upward, and cover them with two inches of earth. This, he says, advances their growth by one year. Possibly, with these hints, we may devise a way to make cuttings from Norton's Virginia Seedling take root, a thing as yet found impossible unless forcing of some kind is resorted to.

After establishing the points to be occupied by the souches in the new plantation, a peg nine inches long is driven down at each point. At the side of each peg a hole is dug measuring twelve inches every way, and so close up to the peg that it makes its appearance midway in the side of the hole (see plate No. 1). The bottom of the hole is covered with two inches of surface soil, but no manure. The peg being now removed, the plant or cutting is made to take its place, with the lower end, however, to the length of two joints, including two eyes, curved to an inclination of forty-five degrees, the base penetrating to the bottom of the two-inch layer of loose soil, as seen in plate No. 2. The hole is now filled up with surface soil, and well packed down.

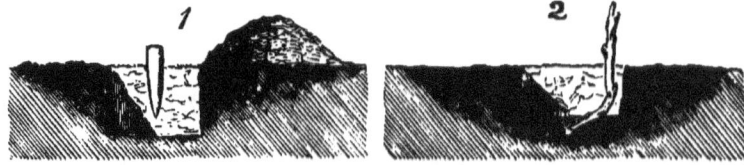

At the proper time, that portion which shows itself above ground is cut back to either two eyes or three; or, if more of it is left to guide the plowman, the buds on it are rubbed off except the two or three lowest. If the soil be damp, three eyes are left, which will give the souche a height of about eight inches; but if the ground be dry, only two are left, giving a height of from four to six inches. For the longer souche a small stake is provided to hold it up till it can maintain itself. Eight or ten inches would be the proper length in this country.

During the first season the ground is worked at least three times, and, if needed, as many as six times, for no weeds must be tolerated. At the second plowing of the second year the earth is removed from the feet of the plants, so as to leave each row standing in a shallow trench, which is not closed until early in May. This is to allow the suckers, which sprout from the plant quite down to its foot, and prevent its forming a good souche, to be removed, which is done just before the trenches are filled, at which time the suckers have usually attained a length of twelve inches.

As the winds in the south of France are extremely violent, a mound of earth four inches high is formed about the plant during the first and second seasons after filling the trenches, otherwise they would be in danger of being uprooted. The second year the cul-

O

tivation is otherwise the same as during the first year.

The third year the same treatment as to suckering and banking up the earth, as well as plowing, is followed.

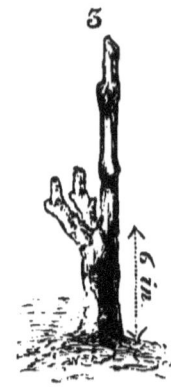

At the end of the first year, if the growth has been good, and two large enough shoots can be found issuing from the proper place, both of them are reserved to form arms to the future souche, and are cut back to two or three eyes, as in plate No. 3.

Pruning in the second and third years is not done till late in March or early in April, lest the young plants, being earlier to put out than older ones, should suffer from frosts.

Pruning at the end of the first year.

At the end of the second year the pruning is so performed as to give to the souche three, four, or six arms, according to its vigor; and, if the soil and cultivation are good, each arm is allowed to retain three eyes, otherwise each is cut back to only two.

Pruning at the end of the second year.

At the end of the third year the souche is so

pruned as to increase the number of arms to six, if that could not be done the year before; if the ground is good, two eyes are this year allowed to each shoot; if not, then only one eye and a sub-eye are reserved.

Pruning at the end of the third year.

We have now conducted the young plant to its adult age, and henceforth the pruning is uniform year by year. This is so conducted as to leave at the extremity of every arm one shoot of the preceding year's growth, cut back to one or two eyes, according to the vigor of the vine. If the vine is very strong, sometimes more than one shoot is left on an arm. Care is taken to balance the souche on all sides by keeping the arms as equal in length and regular in position as possible.

If too much old wood has accumulated on the arms, so as to impair the health of the souche, it is carefully corrected by pruning in the way shown in the two plates, No. 6 and No. 7.

Whenever, from age, disease, or other cause, a vine is condemned to be rooted out, and it becomes good policy to obtain from it all the fruit it will ripen

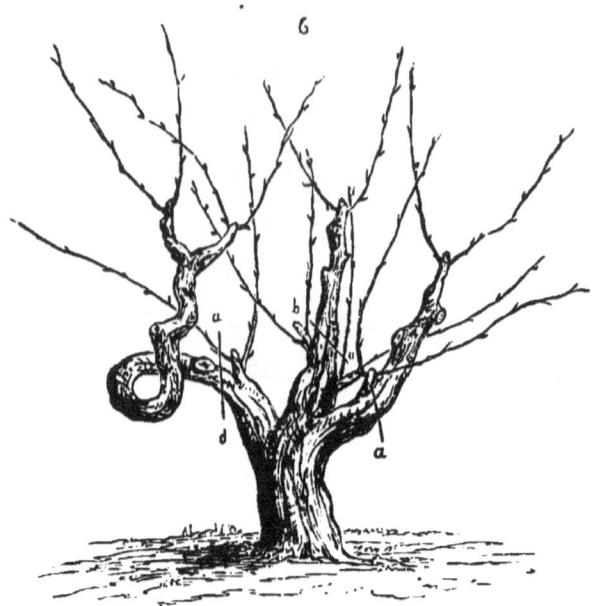

An old souche needing to be reduced.

The same souche after being reduced.

while it lives, two of the shoots are bowed upward and tied together without any pruning, and all the others are cut back to three eyes. This soon exhausts the vine.

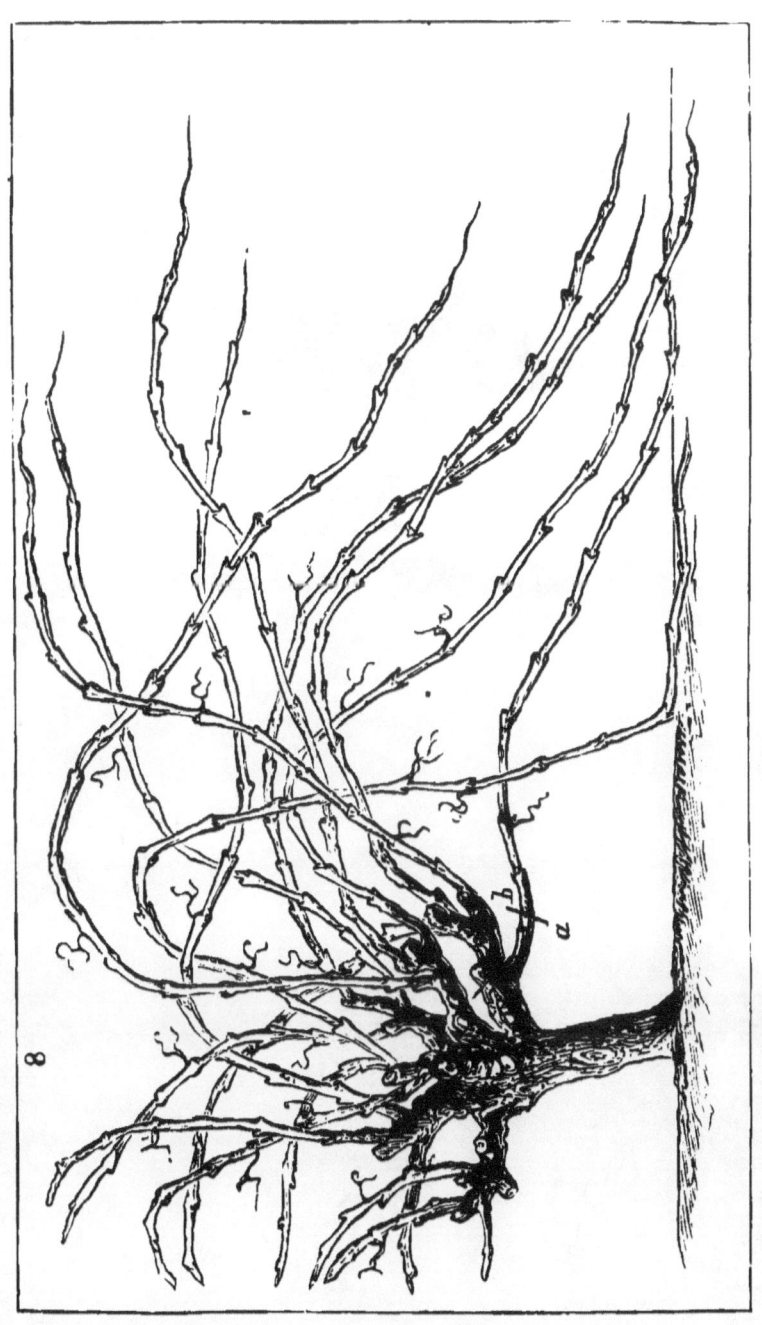

Plate No. 8 represents an Aramon twelve years old.

Plate No. 9 represents a plant forty years old that has been pruned.

CHAPTER XX.

CONCLUSION.

THE climates of those countries from which our knowledge of viticulture has been mainly derived but very little resemble that of the United States. Our skies are clear, our rains torrential, our sun hot, and our air dry. We have rude winds and sudden and severe changes. The important vine-district to which I have been just directing attention has skies as clear, a sun as hot, rains as copious, a drier air, ruder winds, and quicker changes. Its inhabitants enjoy an experience that has come down from Roman and even Grecian days; and a people who but lately led European civilization are certainly able to profit by such an experience, and may safely be presumed to know something about grape-growing.

I commend especially to gentlemen in the Southern States the subject of training according to the modes of Southern France, as being adapted to many of their warmer soils and their abundant sunshine.

How far to the northward of the Valley of the Ohio the method in question might hope for success depends on many facts and considerations. We have seen that the Folle-blanche, thus trained, does perfectly well in a colder climate than that of our Lake Erie shore, and a damper than any on our Atlantic sea-board, while, as for those other plants which flourish throughout Languedoc, they would find as warm a home on the banks of "La Belle Rivière.".

I conclude with a few words of advice to such of my countrymen as can command a half acre of ground for vine-growing. Drain it well, and keep it only moderately rich. Plant Norton's Seedling or Ives's, and train them on the low souche system, following closely its simple precepts, except where your own judgment modifies them. Gather the grapes before they get too ripe, crush them after stemming, and let the must complete its fermentation before being drawn off the skins, so that your wine shall be thoroughly red. Drink that wine—you, and your wife, and little ones; drink it for breakfast, drink it for dinner, drink it for supper; drink it, in short, whenever you are dry, or wet, or cold, or tired. Drink your own wine, and not another's. It will cost less than the sugar you now mix in your tea or coffee. It will reduce your grocer's bill, and nearly abolish your doctor's. It will give you healthy children, and not only purify their blood, but mend their manners also, Babrius says. Your wife will gradually reform her ice-water drinking, abandon her dyspepsy, take flesh upon her bones, be seen to smile often and sometimes laugh, and glow with the warmth of health and love. You, on your part, will gradually become less uneasy, and more fond of amusement than excitement; will grow more plump, but have better strength to carry your flesh. Owing to the presence of particles of red coloring matter in the wine, things will look more rosy-hued than before, and the future appear not unlike a Western sunset. At times you may feel somewhat "above par"—a trifle lighter-hearted than usual. In such case, be not alarmed; excessive happiness is a symptom that will generally pass away of itself—or they might pump a little cold water on your head.

Thus can you obtain in abundance a purer drink than water, a cheaper drink than sugared water, and a healthier one than any. Thus may you bring tranquillity and cheerfulness beneath your roof-tree, and contentment and affection to your fireside—live a merry life, and

"Die a good old man."

O 2

INDEX.

A.

B.

C.

M.

N.

O.

P.

V.

W.

THE END.

VALUABLE STANDARD WORKS

FOR PUBLIC AND PRIVATE LIBRARIES,

PUBLISHED BY HARPER & BROTHERS, NEW YORK.

---

*For a full List of Books suitable for Libraries, see HARPER & BROTHERS'
TRADE-LIST and CATALOGUE, which may be had gratuitously on application
to the Publishers personally, or by letter enclosing Five Cents.*

HARPER & BROTHERS *will send any of the following works by mail, postage
prepaid, to any part of the United States, on receipt of the price.*

MOTLEY'S DUTCH REPUBLIC. The Rise of the Dutch Republic. A
History. By JOHN LOTHROP MOTLEY, LL.D., D.C.L. With a Portrait of
William of Orange. 3 vols., 8vo, Cloth, $10 50.

MOTLEY'S UNITED NETHERLANDS. History of the United Netherlands: from the Death of William the Silent to the Twelve Years' Truce
—1609. With a full View of the English-Dutch Struggle against Spain,
and of the Origin and Destruction of the Spanish Armada. By JOHN
LOTHROP MOTLEY, LL.D., D.C.L., Author of "The Rise of the Dutch Republic." Portraits. 4 vols., 8vo, Cloth, $14 00.

ABBOTT'S LIFE OF CHRIST. Jesus of Nazareth : his Life and Teachings;
Founded on the Four Gospels, and Illustrated by Reference to the Man-
ners, Customs, Religious Beliefs, and Political Institutions of his Times.
By LYMAN ABBOTT. With Designs by Doré, De Laroche, Fenn, and others.
Crown 8vo, Cloth, Beveled Edges, $3 50.

NAPOLEON'S LIFE OF CÆSAR. The History of Julius Cæsar. By His
Imperial Majesty NAPOLEON III. Volumes I. and II. now ready. Library
Edition, 8vo, Cloth, $3 50 per vol.
Maps to Vols. I. and II. sold separately. Price $1 50 each, NET.

HENRY WARD BEECHER'S SERMONS. Sermons by HENRY WARD
BEECHER, Plymouth Church, Brooklyn. Selected from Published and Un-
published Discourses, and Revised by their Author. With Steel Portrait
by Halpin. Complete in Two Vols., 8vo, Cloth, $5 00.

LYMAN BEECHER'S AUTOBIOGRAPHY, &c. Autobiography, Corre-
spondence, &c., of Lyman Beecher, D.D. Edited by his Son, CHARLES
BEECHER. With Three Steel Portraits, and Engravings on Wood. In Two
Vols., 12mo, Cloth, $5 00.

BALDWIN'S PRE-HISTORIC NATIONS. Pre-Historic Nations: or, Inquiries concerning some of the Great Peoples and Civilizations of Antiquity, and their Probable Relation to a still Older Civilization of the Ethi-
opians or Cushites of Arabia. By JOHN D. BALDWIN, Member of the
American Oriental Society. 12mo, Cloth, $1 75.

WHYMPER'S ALASKA. Travel and Adventure in the Territory of Alaska,
formerly Russian America—now Ceded to the United States—and in various other parts of the North Pacific. By FREDERICK WHYMPER. With Map
and Illustrations. Crown 8vo, Cloth, $2 50.

DILKE'S GREATER BRITAIN. Greater Britain: a Record of Travel in
English-speaking Countries during 1866 and 1867. By CHARLES WENT-
WORTH DILKE. With Maps and Illustrations. 12mo, Cloth, $1 00.

LOSSING'S FIELD-BOOK OF THE WAR OF 1812. Pictorial Field-Book of the War of 1812; or, Illustrations, by Pen and Pencil, of the History, Biography, Scenery, Relics, and Traditions of the Last War for American Independence. By BENSON J. LOSSING. With several hundred Engravings on Wood, by Lossing and Barritt, chiefly from Original Sketches by the Author. 1088 pages, 8vo, Cloth, $7 00.

LOSSING'S FIELD-BOOK OF THE REVOLUTION. Pictorial Field-Book of the Revolution; or, Illustrations, by Pen and Pencil, of the History, Biography, Scenery, Relics, and Traditions of the War for Independence. By BENSON J. LOSSING. 2 vols., 8vo, Cloth, $14 00; Sheep, $15 00; Half Calf, $18 00; Full Turkey Morocco, $22 00.

SMILES'S SELF-HELP. Self-Help; with Illustrations of Character and Conduct. By SAMUEL SMILES. 12mo, Cloth, $1 25.

SMILES'S HISTORY OF THE HUGUENOTS. The Huguenots: their Settlements, Churches, and Industries in England and Ireland. By SAMUEL SMILES, Author of "Self-Help," &c. With an Appendix relating to the Huguenots in America. Crown 8vo, Cloth, Beveled, $1 75.

WHITE'S MASSACRE OF ST. BARTHOLOMEW. The Massacre of St. Bartholomew: Preceded by a History of the Religious Wars in the Reign of Charles IX. By HENRY WHITE, M.A. With Illustrations. 8vo, Cloth, $1 75.

ABBOTT'S HISTORY OF THE FRENCH REVOLUTION. The French Revolution of 1789, as viewed in the Light of Republican Institutions. By JOHN S. C. ABBOTT. With 100 Engravings. 8vo, Cloth, $5 00.

ABBOTT'S NAPOLEON BONAPARTE. The History of Napoleon Bonaparte. By JOHN S. C. ABBOTT. With Maps, Woodcuts, and Portraits on Steel. 2 vols., 8vo, Cloth, $10 00.

ABBOTT'S NAPOLEON AT ST. HELENA; or, Interesting Anecdotes and Remarkable Conversations of the Emperor during the Five and a Half Years of his Captivity. Collected from the Memorials of Las Casas, O'Meara, Montholon, Antommarchi, and others. By JOHN S. C. ABBOTT. With Illustrations. 8vo, Cloth, $5 00.

ADDISON'S COMPLETE WORKS. The Works of Joseph Addison, embracing the whole of the "Spectator." Complete in 3 vols., 8vo, Cloth, $6 00.

ALCOCK'S JAPAN. The Capital of the Tycoon: a Narrative of a Three Years' Residence in Japan. By Sir RUTHERFORD ALCOCK, K.C.B., Her Majesty's Envoy Extraordinary and Minister Plenipotentiary in Japan. With Maps and Engravings. 2 vols., 12mo, Cloth, $3 50.

ALFORD'S GREEK TESTAMENT. The Greek Testament: with a critically-revised Text; a Digest of Various Readings; Marginal References to Verbal and Idiomatic Usage; Prolegomena; and a Critical and Exegetical Commentary. For the Use of Theological Students and Ministers. By HENRY ALFORD, D.D., Dean of Canterbury. Vol. I., containing the Four Gospels. 944 pages, 8vo, Cloth, $6 00; Sheep, $6 50.

ALISON'S HISTORY OF EUROPE. FIRST SERIES: From the Commencement of the French Revolution, in 1789, to the Restoration of the Bourbons, in 1815. [In addition to the Notes on Chapter LXXVI., which correct the errors of the original work concerning the United States, a copious Analytical Index has been appended to this American edition.] SECOND SERIES: From the Fall of Napoleon, in 1815, to the Accession of Louis Napoleon, in 1852. 8 vols., 8vo, Cloth, $16 00.

BANCROFT'S MISCELLANIES. Literary and Historical Miscellanies. By GEORGE BANCROFT. 8vo, Cloth, $3 00.

DRAPER'S CIVIL WAR. History of the American Civil War. By JOHN W. DRAPER, M.D., LL.D., Professor of Chemistry and Physiology in the University of New York. In Three Vols. *Vol. II. just published.* 8vo, Cloth, $3 50 per vol.

DRAPER'S INTELLECTUAL DEVELOPMENT OF EUROPE. A History of the Intellectual Development of Europe. By JOHN W. DRAPER, M.D., LL.D., Professor of Chemistry and Physiology in the University of New York. 8vo, Cloth, $5 00.

DRAPER'S AMERICAN CIVIL POLICY. Thoughts on the Future Civil Policy of America. By JOHN W. DRAPER, M.D., LL.D., Professor of Chemistry and Physiology in the University of New York, Author of a "Treatise on Human Physiology," "A History of the Intellectual Development of Europe," &c. Crown 8vo, Cloth, $2 50.

BARTH'S NORTH AND CENTRAL AFRICA. Travels and Discoveries in North and Central Africa: being a Journal of an Expedition undertaken under the Auspices of H.B.M.'s Government, in the Years 1849-1855. By HENRY BARTH, Ph.D., D.C.L. Illustrated. Complete in Three Vols., 8vo, Cloth, $12 00.

BELLOWS'S OLD WORLD. The Old World in its New Face: Impressions of Europe in 1867-1868. By HENRY W. BELLOWS. 2 vols., 12mo, Cloth, $3 50.

BOSWELL'S JOHNSON. The Life of Samuel Johnson, LL.D. Including a Journey to the Hebrides. By JAMES BOSWELL, Esq. A New Edition, with numerous Additions and Notes. By JOHN WILSON CROKER, LL.D., F.R.S. Portrait of Boswell. 2 vols., 8vo, Cloth, $4 00.

BRODHEAD'S HISTORY OF NEW YORK. History of the State of New York. By JOHN ROMEYN BRODHEAD. First Period, 1609-1664. 8vo, Cloth, $3 00.

BULWER'S PROSE WORKS. Miscellaneous Prose Works of Edward Bulwer, Lord Lytton. In Two Vols. 12mo, Cloth, $3 50.

BURNS'S LIFE AND WORKS. The Life and Works of Robert Burns. Edited by ROBERT CHAMBERS. 4 vols., 12mo, Cloth, $6 00.

CARLYLE'S FREDERICK THE GREAT. History of Friedrich II., called Frederick the Great. By THOMAS CARLYLE. Portraits, Maps, Plans, &c. 6 vols., 12mo, Cloth, $12 00.

CARLYLE'S FRENCH REVOLUTION. History of the French Revolution. Newly Revised by the Author, with Index, &c. 2 vols., 12mo, Cloth, $3 50.

CARLYLE'S OLIVER CROMWELL. Letters and Speeches of Oliver Cromwell. With Elucidations and Connecting Narrative. 2 vols., 12mo, Cloth, $3 50.

CHALMERS'S POSTHUMOUS WORKS. The Posthumous Works of Dr. Chalmers. Edited by his Son-in-Law, Rev. WILLIAM HANNA, LL.D. Complete in Nine Vols., 12mo, Cloth, $13 50.

CLAYTON'S QUEENS OF SONG. Queens of Song: being Memoirs of some of the most celebrated Female Vocalists who have performed on the Lyric Stage from the Earliest Days of Opera to the Present Time. To which is added a Chronological List of all the Operas that have been performed in Europe. By ELLEN CREATHORNE CLAYTON. With Portraits. 8vo, Cloth, $3 00.

COLERIDGE'S COMPLETE WORKS. The Complete Works of Samuel Taylor Coleridge. With an Introductory Essay upon his Philosophical and Theological Opinions. Edited by Professor SHEDD. Complete in Seven Vols. With a fine Portrait. Small 8vo, Cloth. $10 50.

4 *Harper & Brothers' Valuable Standard Works.*

DU CHAILLU'S AFRICA. Explorations and Adventures in Equatorial Africa: with Accounts of the Manners and Customs of the People, and of the Chase of the Gorilla, the Crocodile, Leopard, Elephant, Hippopotamus, and other Animals. By PAUL B. DU CHAILLU, Corresponding Member of the American Ethnological Society; of the Geographical and Statistical Society of New York; and of the Boston Society of Natural History. With numerous Illustrations. 8vo, Cloth, $5 00.

DU CHAILLU'S ASHANGO LAND. A Journey to Ashango Land: and Further Penetration into Equatorial Africa. By PAUL B. DU CHAILLU, Author of "Discoveries in Equatorial Africa," &c. New Edition. Handsomely Illustrated. 8vo, Cloth, $5 00.

CURTIS'S HISTORY OF THE CONSTITUTION. History of the Origin, Formation, and Adoption of the Constitution of the United States. By GEORGE TICKNOR CURTIS. Complete in Two large and handsome Octavo Volumes. Cloth, $6 00.

DAVIS'S CARTHAGE. Carthage and her Remains: being an Account of the Excavations and Researches on the Site of the Phœnician Metropolis in Africa and other adjacent Places. Conducted under the Auspices of Her Majesty's Government. By Dr. DAVIS, F.R.G.S. Profusely Illustrated with Maps, Woodcuts, Chromo-Lithographs, &c. 8vo, Cloth, $4 00.

DOOLITTLE'S CHINA. Social Life of the Chinese: with some Account of their Religious, Governmental, Educational, and Business Customs and Opinions. With special but not exclusive Reference to Fuhchau. By Rev. JUSTUS DOOLITTLE, Fourteen Years Member of the Fuhchau Mission of the American Board. Illustrated with more than 150 characteristic Engravings on Wood. 2 vols., 12mo, Cloth, $5 00.

EDGEWORTH'S (Miss) NOVELS. With Engravings. 10 vols., 12mo, Cloth, $15 00.

GIBBON'S ROME. History of the Decline and Fall of the Roman Empire. By EDWARD GIBBON. With Notes by Rev. H. H. MILMAN and M. GUIZOT. A new cheap Edition. To which is added a complete Index of the whole Work, and a Portrait of the Author. 6 vols., 12mo (uniform with Hume), Cloth, $9 00.

GROTE'S HISTORY OF GREECE. 12 vols., 12mo, Cloth, $18 00.

HALE'S (MRS.) WOMAN'S RECORD. Woman's Record; or, Biographical Sketches of all Distinguished Women, from the Creation to the Present Time. Arranged in Four Eras, with Selections from Female Writers of each Era. By Mrs. SARAH JOSEPHA HALE. Illustrated with more than 200 Portraits. 8vo, Cloth, $5 00.

HALL'S ARCTIC RESEARCHES. Arctic Researches and Life among the Esquimaux: being the Narrative of an Expedition in Search of Sir John Franklin, in the Years 1860, 1861, and 1862. By CHARLES FRANCIS HALL. With Maps and 100 Illustrations. The Illustrations are from Original Drawings by Charles Parsons, Henry L. Stephens, Solomon Eytinge, W. S. L. Jewett, and Granville Perkins, after Sketches by Captain Hall. A New Edition. 8vo, Cloth, Beveled Edges, $5 00.

HALLAM'S CONSTITUTIONAL HISTORY OF ENGLAND, from the Accession of Henry VII. to the Death of George II. 8vo, Cloth, $2 00.

HALLAM'S LITERATURE. Introduction to the Literature of Europe during the Fifteenth, Sixteenth, and Seventeenth Centuries. By HENRY HALLAM. 2 vols., 8vo, Cloth, $4 00.

HALLAM'S MIDDLE AGES. State of Europe during the Middle Ages. By HENRY HALLAM. 8vo, Cloth, $2 00.